Graphing Calculator Manual and Student Solutions Manual
to accompany

EXPLORATIONS IN COLLEGE ALGEBRA

Second Edition

Linda Almgren Kime
Judith Clark
University of Massachusetts, Boston

GRAPHING CALCULATOR MANUAL PREPARED BY
Beverly K. Michael
University of Pittsburgh

STUDENT SOLUTIONS MANUAL PREPARED BY
John A. Lutts
University of Massachusetts, Boston

JOHN WILEY & SONS, INC.
NEW YORK • CHICHESTER • WEINHEIM • BRISBANE • SINGAPORE • TORONTO

COVER PHOTO: ©Jeff Hunter/The Image Bank

To order books or for customer service call 1-800-CALL-WILEY (225-5945).

ISBN 0-471-40356-3

Printed in the United States of America.

10 9 8 7 6 5 4 3 2 1

Printed and bound by Hamilton Printing, Inc.

Instruct a wise person and he becomes still wiser;
teach a just person and he advances in learning.
Proverbs 9:9

Part I

Graphing Calculator Manual

PREFACE

To the student

- Each chapter in this *Graphing Calculator Workbook* is designed to correspond with each chapter in *Explorations in College Algebra.* The exception is Chapter 0, which is to get you started.
- There is no need to learn everything at once. Calculator techniques are introduced as you need them in each chapter.
- It is assumed that you are using either a TI-82 or TI-83 Graphing Calculator and that you are working right along with the author.
- Since the text exercises assumes that you have either a computer or graphing calculator available, you will want to keep this manual available as a reference when you are doing the exercises.
- You should work through each chapter in the *Graphing Calculator Workbook* before you attempt to do the text exercises.
- Your instructor may assign the *Graphing Calculator Workbook* chapters for homework, for an in-class activity or for an in-class demonstration. Whatever technique is used, follow along with your own calculator. One will learn techniques best by doing.
- The "rule of three" is followed throughout the *Graphing Calculator Workbook.* Mathematical ideas are introduced graphically, numerically and algebraically. For complete understanding you should be able to solve a mathematical problem in a variety of ways.
- A graphing calculator is a tool. It is only as good as its operator. You should practice using this tool for the right job. One of the best ways to use the calculator is to verify your answer. Another is to perform mathematical experiments by creating "what if" situations.
- Learning anything should be fun. Enjoy yourself. Remember confusion is the first step to learning, if you are not confused then you are at a practice stage and practice makes perfect.

Acknowledgments

Thank you to:

- Linda and Judy for being inspired to write their book.
- Dan, Megan and Bridget for their patience and love.
- Frank and Bert for sharing their enthusiasm for graphing technology.

Beverly K. Michael
bkm@pitt.edu

Part I: Graphing Calculator Manual

Contents

All chapters in this Graphing Calculator Manual are intended to teach the graphing calculator skills necessary to do the homework problems in *Explorations in College Algebra, 2nd Edition* by Linda Kime and Judy Clark. John Wiley and Sons Inc. New York. 2001.

Chapter 0
Getting Started on the TI-82 or TI-83[1]

0.1 Turn the Calculator ON / OFF Locating the keys.

Turn your calculator on by using the [ON] key, located in the lower left hand corner of the calculator. To turn the calculator off press [2nd] [OFF] : located above the [ON] key.

To locate the correct keys think of your calculator as being divided into three sections:

1. The bottom six row of keys are your mathematical calculation and function keys.
2. Rows 7 - 9 are the menu and editing keys.
3. The very top row (under the screen) is where your graphing keys are located.

0.2 Adjusting the Screen Contrast

Depending on the room lighting you may want to adjust the screen contrast.

1. To darken the screen:

Press and release the [2nd] key, then press and hold the up arrow [Δ] key.

2. To lighten the screen:

Press and release the [2nd] key, then press and hold the down arrow [∇] key.

As the display contrast changes, a number appears in the upper right corner of the screen between 0 (lightest) and 9 (darkest).

If you adjust the setting to 0, the display may become completely blank. If this happens, increase the contrast and the display will reappear. When contrast needs to be set at 8 or 9 all the time, it is probably time to change the batteries.

0.3 MODE Default Settings

The calculator should be set to the default mode settings. Press [MODE] to see the settings.

Set your calculator to the settings in Figure 0.1 or 0.2 using your arrow keys and pressing [ENTER] to activate your choice.

Figure 0.1

The TI-82 default mode screen.

Figure 0.2

The TI-83 default mode screen.

<u>Note</u>: If your calculator is not new you may want to RESET MEMORY. This will completely erase all data and programs and reset the calculator to the default mode. Use this cautiously. Press [2nd] [MEM] (above +),select **3** then select **[2:Reset]**.

---

[1] The key stroking and menus for the TI-82 and TI-83 are nearly the same. Where they differ, screen images are presented side by side.

0.4 The Home Screen

The Home Screen is your calculation and execution of instruction screen. To return to the Home Screen from any other screen, press [2nd] [QUIT]. The Home Screen is the primary screen of the TI-82 or TI-83.
If there is something displayed on the Home Screen, press the [CLEAR] key.

0.5 Calculating

The bottom six rows of keys on the graphing calculator behave like those on any scientific calculator, except that your entry is seen on an eight line computer screen. When you want the calculator to perform any calculation or instruction, press [ENTER].

> **Note**: The [2nd] key will access the commands to the above left of any key, which are color coded with [2nd] key.

Example 1

$12 \times 2 = ?$

From the Home screen, do the following:

1. Type 12 [X] 2 then press [ENTER]; 24 is now displayed and *stored* as the answer. See Figure 0.3.
2. Press [2nd] [ANS] and [ENTER]; 24 is again displayed.

> **Note**: The result of your last calculation is always stored in memory. To recall your last calculation press [2nd] [ANS].

3. Press the multiplication key [X], then 2 and then [ENTER]. Pressing any operation key, +, -, x, ÷, x^2, x^{-1} etc., assumes that you want to operate on the stored answer. See Figure 0.3.

```
12*2
                24
Ans
                24
Ans*2
                48
                96
```

Figure 0.3

The asterisk, *, is used for multiplication in place of the "times" sign to avoid confusion with the letter x.

0.6 Iteration, Recalling a Process

Notice how [ANS] is also used.

Repeatedly press [ENTER]. Your screen should look like the bottom of Figures 0.3 and 0.4. This process is called <u>iteration</u> (repeating some process over and over again). The last operation (multiplying by 2) is repeated on the new answer.

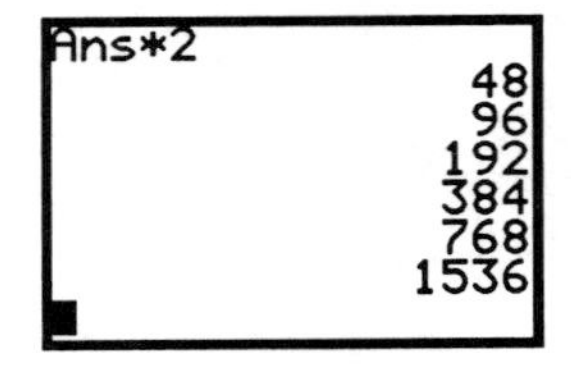

Figure 0.4

Example 2

Interest compounded at 5% annually on an initial investment of $1000 can be represented by 1000*1.05, or A = P(1 + R) for the first year .
[Amount = (original investment)(1 + rate).]
Use iteration to determine the number of years for the amount of accumulated investment to be greater than $1300.

Press [CLEAR] to clear the Home Screen.

Type 1000 followed by [ENTER] .

The number 1000 is now stored in memory.

Press [X] 1.05 [ENTER] . The number 1050 will now be displayed. By repeatedly pressing [ENTER] , you can see the growth of your initial $1000 investment year by year and determine that 6 iterations (years) are necessary for you to exceed $1300. See Figures 0.5 and 0.6.

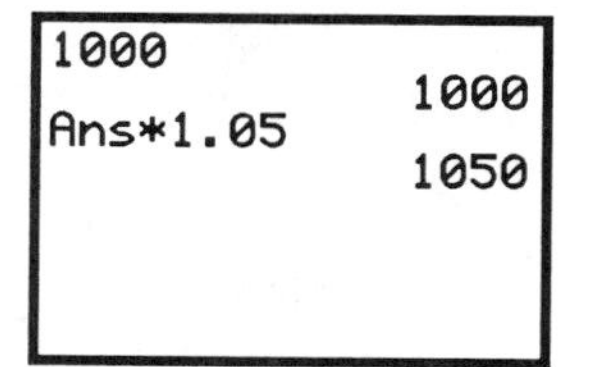

Figure 0.5

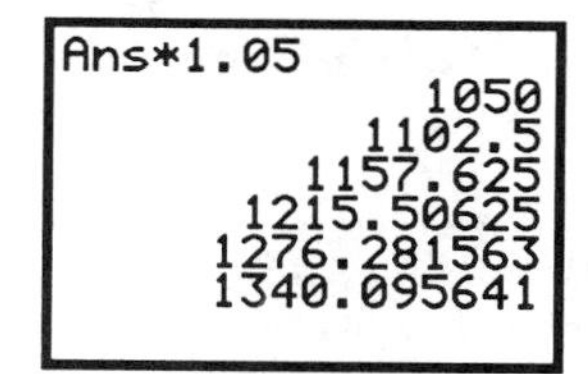

Figure 0.6

Between year 5 and 6 the amount is > 1300.

0.7 Converting Decimals and Fractions

The TI-82 or TI-83 can be used to convert decimals and fractions.

Press 1 [÷] 4 [ENTER] .See Figure 0.7.

The decimal answer for this expression, .25, is displayed. Press [MATH] . You are in the MATH menu . Menus give a list of additional command options. See Figure 0. 8 or 0.9. Press [1] or [ENTER] to select the highlighted option. This option [1: Frac] will change the decimal answer back into a fraction.

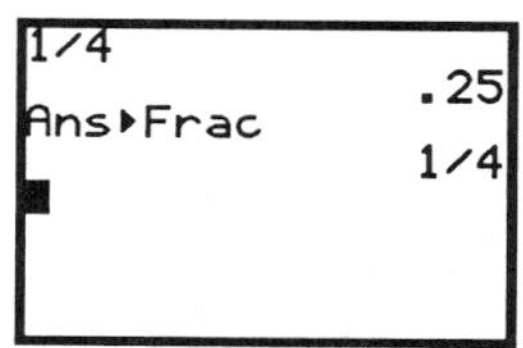

Figure 0.7

> <u>Note</u>: When the denominator of a fraction has more than four digits the answer is displayed as a decimal and will not return to a fraction.

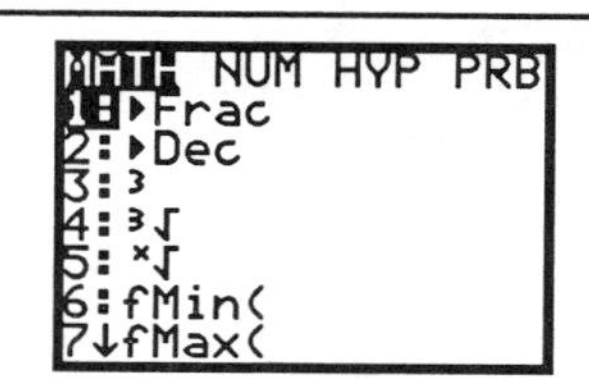

Figure 0.8

The TI-82 MATH menu

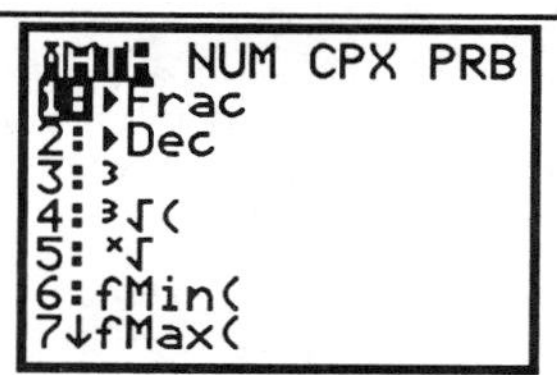

Figure 0.9

The TI-83 MATH menu

0.8 Selecting Items from a Menu.

You can select an item from a menu by typing the number or by moving to that menu option with the down arrow key ∇ .

You press ENTER to select your menu option. Press MATH ∇ select [2: Dec] .

Press ENTER , this changes the fraction back to a decimal. See Figure 0.10.

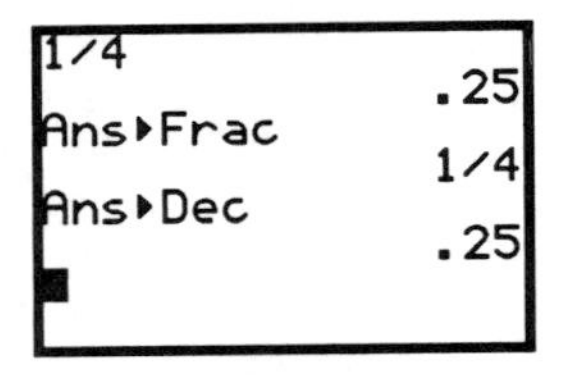

Figure 0.10

Example 3

Type in the following fraction problems, then use the MATH menu to change the answers back to fractional form.

1. $\frac{1}{2}+\frac{1}{3}$

   Press (1 ÷ 2) + (1 ÷) 3 ENTER .

   Press MATH 1 ENTER .

2. $3\frac{5}{9}+5\frac{3}{7}$

   Press (3 + 5 ÷ 9) + (5 + 3 ÷ 7)

   ENTER . Press MATH 1 ENTER

   See Figure 0.11.

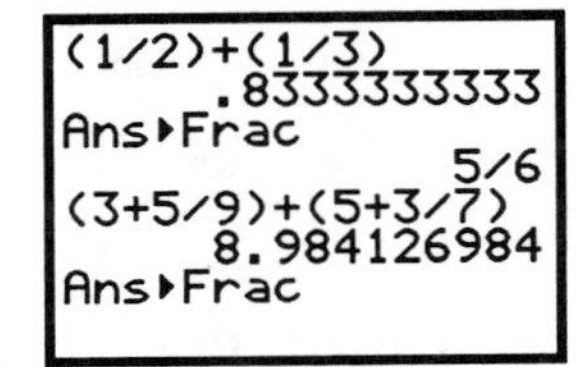

Figure 0.11

The answer to Example 3 part 2 is 566/63.

Press CLEAR .

0.9 Raising a Number to a Power

The calculator can be used to raise a number (called the base) to a power by using the exponent key, ^ .

For 3^2 press 3 ^ 2 ENTER or use a short - cut, press 3 x^2 ENTER . This last method pastes the exponent to the right of 3. See Figure 0.12.

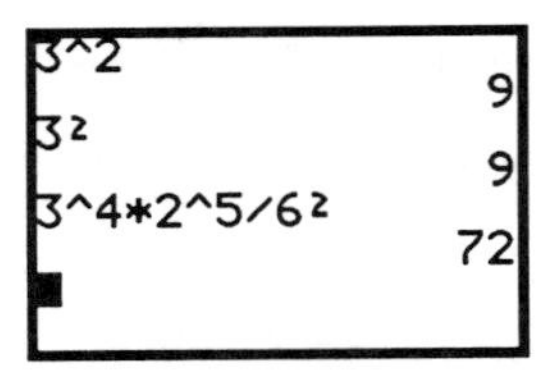

Figure 0.12

Example 4

Type the expression:

3^4 x $2^5 \div 6^2$, then press ENTER .

For the result see Figure 0 12.

0.10 Order of Operations

The TI-82 or TI-83 uses an algebraic order of operations: inside parentheses first, powers next, then multiply or divide from left to right and lastly add or subtract from left to right.

Example 5

1. **Enter: $1 + 2(4 - 2)^2 + 6 \div 2$**
   See Figure 13.

The order of operations are performed algebraically in the following steps:

$1 + 2(4 - 2)^2 + 6 \div 2 =$	
$1 + 2(\mathbf{2})^2 + 6 \div 2 =$	inside parentheses
$1 + 2(\mathbf{4}) + 6 \div 2 =$	raise to power
$1 + \mathbf{8} + 6 \div 2 =$	multiply
$1 + 8 + \mathbf{3} =$	divide
$\mathbf{9} + 3 =$	add
$\mathbf{12}$	add

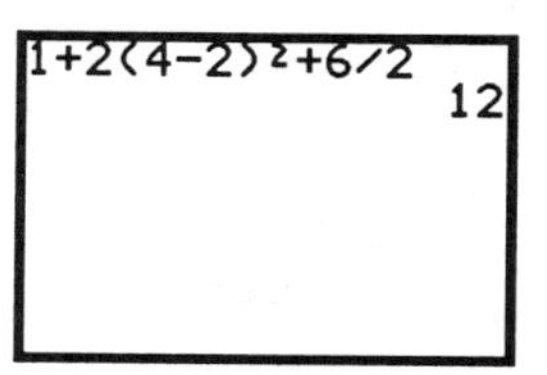

Figure 0.13

2. **Enter: "One hundred fifths times two."**
   See Figure 0.14 for two methods.

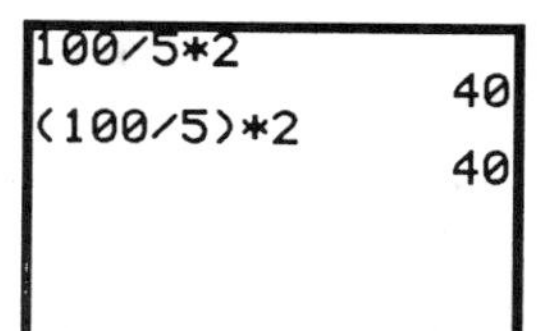

Figure 0.14

Troubleshooting: For the TI-82 users, compare the difference in order of operations in Figures 0.14 and 0.15. Parenthesis in the denominator of a fraction are interpreted as a grouping .
For the TI-83 parentheses are interpreted the same as the multiplication sign. See Figure 0.16.
To avoid confusion always enclose fractions in parentheses.

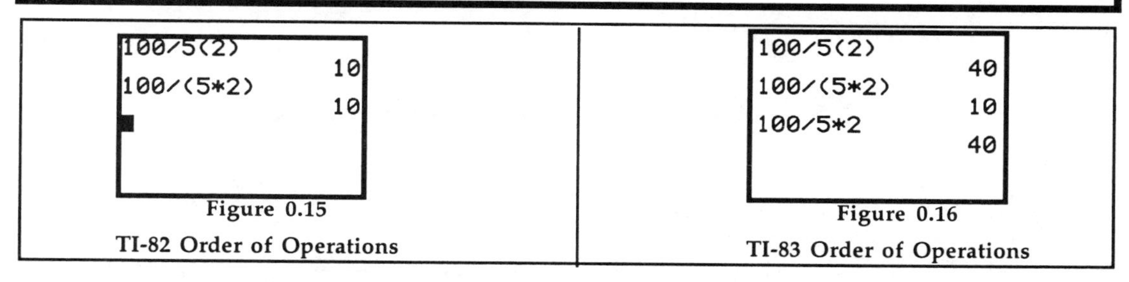

Figure 0.15
TI-82 Order of Operations

Figure 0.16
TI-83 Order of Operations

3. **Enter "Sixteen raised to the one half power."**

This is the same as "the square root of 16."

Note: 16^1/2 is ***not*** √16; fractional exponents must always be enclosed in parentheses. See Figure 0.17.

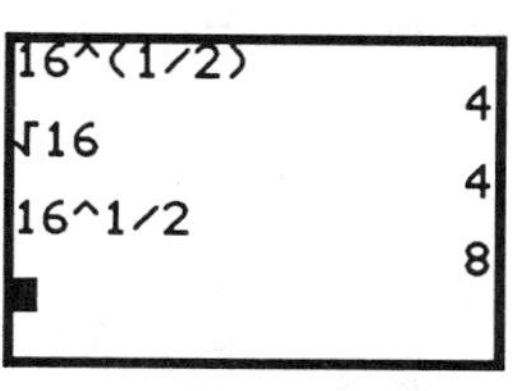

Figure 0.17

0.11 Truth Tests

The graphing calculator can be used to determine whether an expression is true or false.

To use this feature, you must use the [2nd] [TEST] menu. Figure 0.18 shows the TEST menu. This is where the equal and inequality symbols are located.

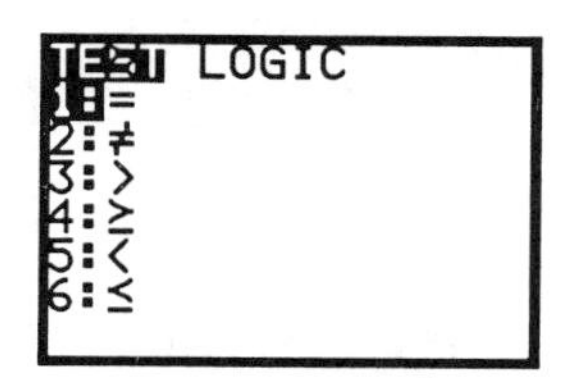

Figure 0.18

Example 6

1. Is $3 < 7$ true or false?

Press 3 [2nd] [TEST] [▽].

Select [5:<] ,press 7 [ENTER]. See Figure 0.19.

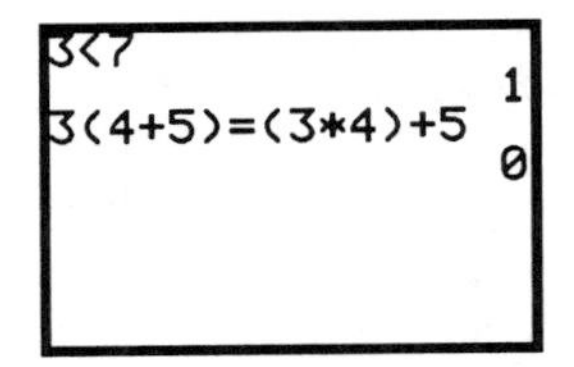

Figure 0.19

> **Note**: When performing a TEST, remember that 1 means TRUE and 0 means FALSE.

2. Is 3 (4 + 5) = (3 x 4) + 5 true or false?

This is a false statement, thus the answer is zero, see Figure 0.19, because:

3(4 + 5) = (3 x 4) + (3 x 5)

0.12 Deep Recall and Editing

Press [CLEAR]. To recover your last entry press [2nd] [ENTRY]. To evaluate press [ENTER]. To edit an expression, use the left and right arrows to position the cursor for editing and press delete [DEL] or insert [2nd] [INS].

Example 7

Change the expression in Example 6 part 2 to: 3(4 + 5) = (3 x 4) + (3 x 5).

First recall the expression, [2nd] [ENTRY].

Use [◁] to place the cursor on the 5; press [2nd] [INS] type [(] 3 [X]. Then [▷] to place the parentheses after the 5. Press [ENTER]. See Figure 0.20. Now the expression is evaluated as true (i.e. the number 1).

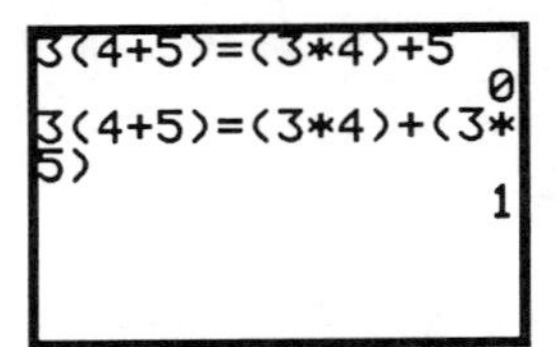

Figure 0.20

> **Note:** Also try pressing [2nd] [ENTRY] several times and you will see some of the old expressions you typed. This is called ***deep recall*** and it is used to retrieve expressions that have been typed many steps earlier.

0.13 Storing Values to Variables

Recall Example 2 where we were finding the amount of money accumulated after one year (A) using the formula $A = P(1 + R)^x$, where the principle P = \$1000 and the rate R = 5%, and x=1. The calculator allows you to store values to alphabetical letters A through Z. You access the letters by first pressing the ALPHA key and you store number values to letters by using the store STO key.

> Note: Alphabetical letters are located to the above right of keys and are color coded to match the ALPHA key.

Example 8

Find A if $P = 1000$, $R = .05$ and $x = 1$ using
$A = P(1 + R)^1 = P(1 + R)$.

1. To store 1000 to P, press 1000 STO ALPHA P ENTER .
2. To store .05 to R, press .05 STO ALPHA R ENTER See Figure 0. 21.
3. Type the expression $P(1 + R)$; remember to press ALPHA before typing the letter. Press ENTER to evaluate.

```
1000→P
                1000
.05→R
                 .05
P(1+R)
                1050
```

Figure 0.21

The expression has been evaluated using the stored values to P and R. These values will remain the same until you store a new value to R and P. See Figure 0. 21.

> Trouble Shooting: If your calculator is new or if the memory has been cleared, the initial stored value to all letters is zero.

0.13.1 A special note about x and y

Since the variables x and y are used in plotting graphs, their values are constantly updated when you TRACE on a graph. Therefore the values of x and y may change if you have used the graphing feature.

There are two ways to access the x variable since it is usually the variable of choice in algebra. Press ALPHA X or use the handy X T Θ key. See Figure 0.22.

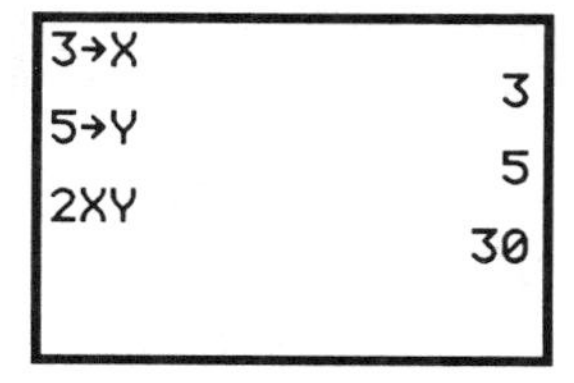

Figure 0.22

0.14 Subtraction and "Negative of"

In algebra the minus sign is used two different ways:

1. as the operation sign between two numbers to mean "subtract", as in 5 - 3 or,
2. in front of a number to mean "the opposite of or negative of", as in -7.

The calculator has two different keys for minus. Press 5 [-] 3 [ENTER] for subtraction.

For ⁻7 find the negative key [(-)] located to the left of ENTER . Press [(-)] 7 [ENTER]

See Figure. 0.23.

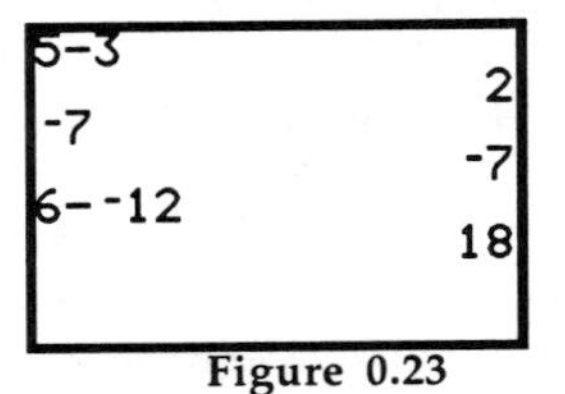

Figure 0.23

> **Note**: The negative sign is actually a little bit shorter and slightly raised compared to the subtraction symbol.

Example 9

Type the following problems:

a) 6 - ⁻ 12

b) - 3 x - 9

c) $(-5)^2$

d) -5^2

See Figures. 0.23 and 0.24.

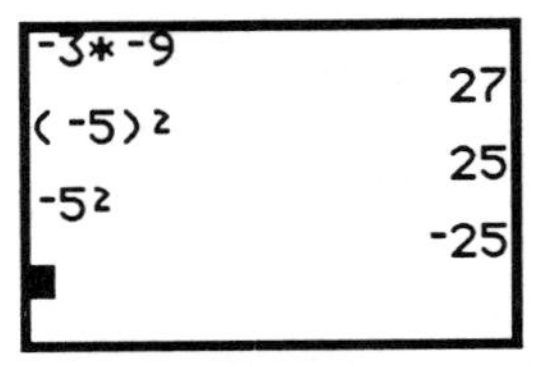

Figure 0.24

Note that the values for example **9c** and **9d** above are different. See Figure 0.24. Order of operations in **9d** says: "Square five then take its opposite."

> **Note:** To square a negative number you must put it in parentheses.

> **Trouble Shooting:** The most common calculator error is using the subtraction symbol instead of the negative symbol. See Figure. 0. 25 and 0.26.

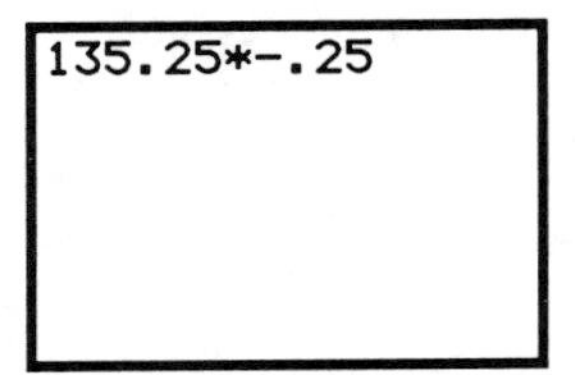

Figure 0.25

0.15 The Error Message

Using the subtraction sign incorrectly produces an error message. When you type the expression as in Figure 0.25 and press [ENTER] **ERR:SYNTAX** appears. See Figure 0.26. Choose [2:Goto] to position the cursor to the place where the error occurred. Choose [1:Quit] to begin a new line on the Home Screen. See Figure 0.26.

Figure 0.26

Chapter 1
Histograms, Scatterplots and Graphs of Functions

1.1 Using Lists for Data Entry

To enter data into the calculator you use the statistics menu. You can store data into lists labeled L1 through L6.

> **Note:** The TI-83 can store data to additional lists by giving them a name. See the owners manual for more detail.

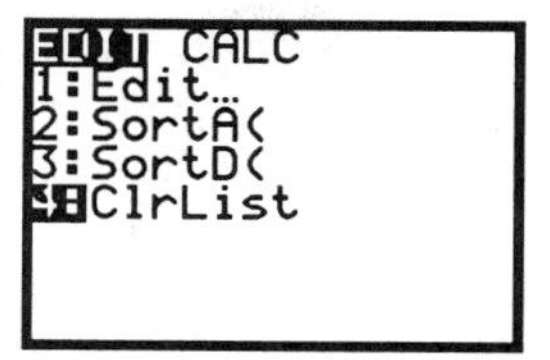

Figure 1. 1

Press [STAT] . To clear any data stored in List 1, and List 2, select [4:ClrList] then press [2nd] [L1] [,] [2nd] [L2] [ENTER]
See Figures. 1.1 and 1.2.

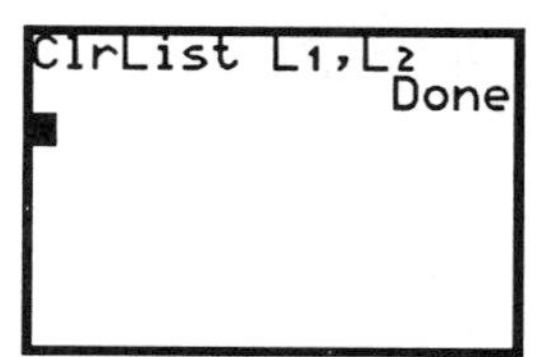

Figure 1. 2

Now you are ready for data entry.

Press [STAT] ; select [1:Edit] See Figure 1.3.

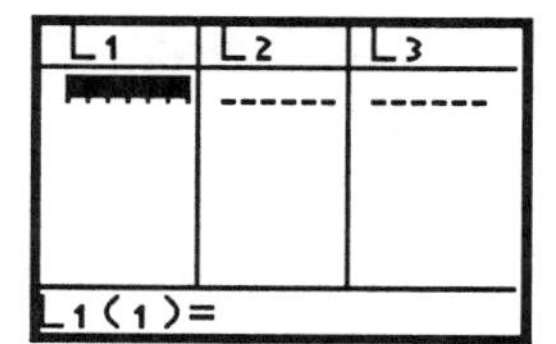

Figure 1. 3

Example 1

Ten students were surveyed. Make a histogram for the number of hours the students worked last week :

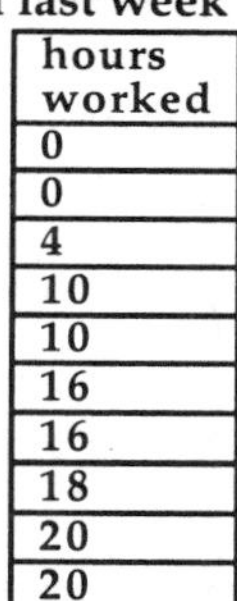

hours worked
0
0
4
10
10
16
16
18
20
20

Table 1. 1

Enter the above data into List 1 (L1).
In column L1 type [0] [ENTER] [0] [ENTER] [4] [ENTER] ... etc. Always press [ENTER] or [▽] after each data entry . See Figure 1.4.

Figure 1. 4

> **Trouble Shooting:** Placing the cursor on the label L1 will display the list using braces { }. If you press [CLEAR] the list will be deleted. This is a quick way to clear. When the down arrow is pressed you can begin entering data into the list.

1. 2 Histogram Setup

A histogram is a graph that helps you visualize data. To graph the data in Table 1.1, first set up the statistical plot.

Clear all old plots. Press [2nd] [STAT PLOT] select [4:PlotsOff] [ENTER]. Press [2nd] [STAT PLOT]; select [1:Plot1] as shown in Figure 1.5. To set up the histogram correctly see Figure 1.6 or 1.7 or follow the instructions below:

1. Select **ON**. Press [ENTER].
2. Select the graph **Type:** [▽] [▷] ...to the **histogram icon**. Press [ENTER].
3. Select the **Xlist: L1**. TI-83 users press [2nd] [L1]. Press [ENTER]

<u>Note</u>: For the TI-83, you cannot type L1 you must press [2nd] [L1].

Select the frequency, **Freq**. In Example 1 each item has a frequency of one so select [1] Press [ENTER]. Your plot should be set up as in Figure 1.6 or 1.7.

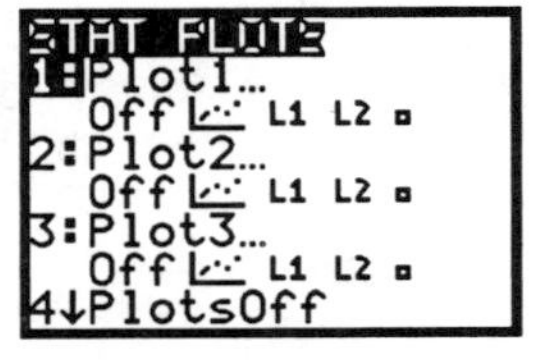

Figure 1. 5

Plot1
On Off
Type:
Xlist: L1 L2 L3 L4 L5 L6
Freq: 1 L1 L2 L3 L4 L5 L6

Figure 1. 6 TI-82 Plot1 menu.

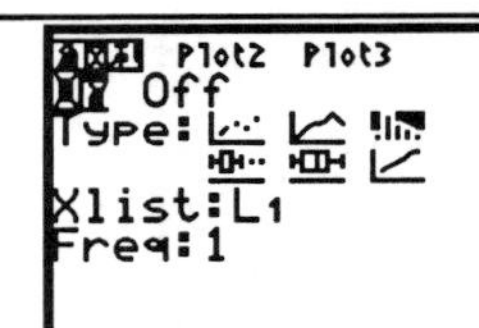

Figure 1. 7 TI-83 Plot1 menu.

1.3 Selecting the Correct Window for the Histogram

<u>Note</u>: **Before you plot a STAT PLOT graph ,**

1. Clear [Y=] or turn off all graphs. To turn off graphs, place the cursor on the = sign then press [ENTER].

2. Turn OFF all plots except the one you want to see.

<u>ERR: INVALID DIM</u> means your lists are not the same size, you have selected a list with no data in it or you have a plot turned on that you did not want and it has different size lists.

1.3.1 Manual Window Sizing

To see your histogram you must set the graphing calculator window to the correct size. Manually set your window size based on the data in L1. Since the L1 data from Example 1 starts at 0 and ends at 20, your x **minimum** and x **maximum** values must be at least: 0 through 20. X maximum should be slightly larger than the highest data value. Y maximum should be slightly larger than the highest frequency of any data entry. Setting the y boundary values at -3 to 5 allows you to see the x-axis as well as the largest frequency value.

Press [WINDOW] [▽] and enter the window settings as in Figure 1.8.

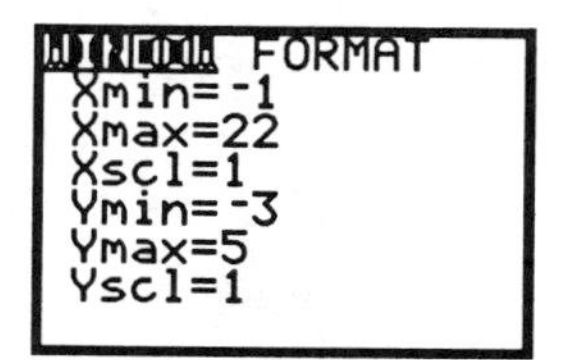

Figure 1. 8

Press [GRAPH] then [TRACE] . Use right arrow [▷] until your screen looks like Figure 1.9.

When you TRACE on a histogram, the cursor moves to the top center of each column. The P1 in the upper right corner indicates you are tracing Plot 1. The **min** = 16, **max** = 1 7 indicates x is on the interval 16 to 17 or, in this case 16 hours. The n value (frequency) is 2. So there are two people who worked 16 hours last week.

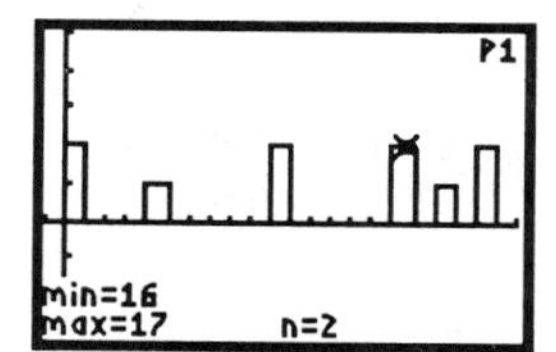

Figure 1. 9

1.4 Working with More than One List

Example 2

Graph a histogram with two data lists.

List 1 represents the sum of the numbers on a pair of dice. List2 represents the possible ways of getting each number by tossing a pair of dice (the frequency). Since we already have data in L1 , enter the data from Table 1.2 below into List 2 (L2) and List 3 (L3).

Clear any data stored in List 2, and List 3.

Press [STAT] ; select [4:ClrList]. Press [2nd] [L2] [,] [2nd] [L3] [ENTER] . See Figures 1.10 and 1.11).

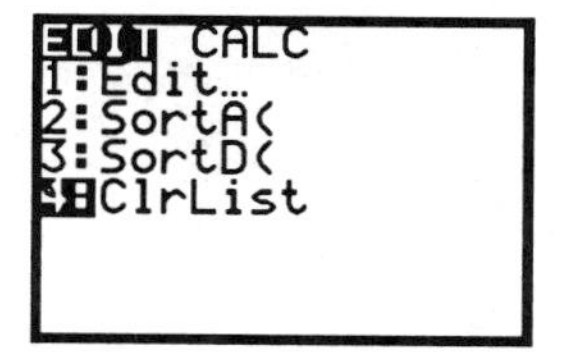

Figure 1. 10

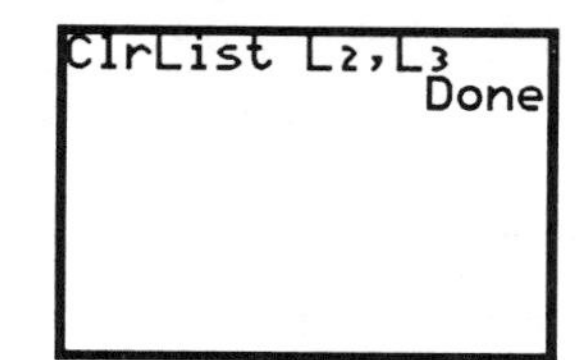

Figure 1. 11

L2	L3
2	1
3	2
4	3
5	4
6	5
7	6
8	5
9	4
10	3
11	2
12	1

Table 1. 2

> **Trouble Shooting:** When using lists as frequency, the data must be whole numbers. Decimal values for frequency will produce the error message **ERR:STAT** on the TI-82.

1.4.1 Entering the Data

Press [STAT] ; select [1:Edit]. See Figure 1.12. Type in the L2 data from Table 1.2. Press 2 [ENTER] 3 [ENTER] 4 [ENTER] ... etc.; always press [ENTER] or [▽] after each data entry. See Figure 1.13.

Press [▷] to get to column L3. Type in all the L3 data from Table 1.2.

Press [ENTER] or [▽] after each number. See Figure 1.13.

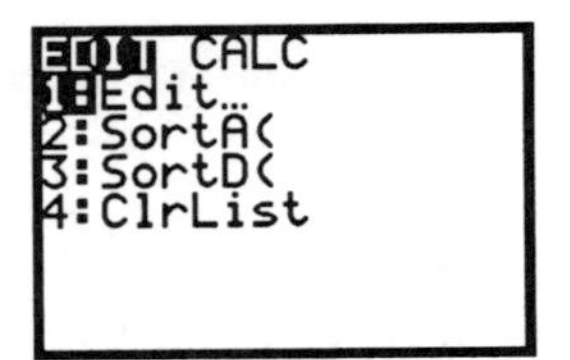

Figure 1. 12

L1	L2	L3
0	2	1
0	3	2
4	4	3
10	5	4
10	6	5
16	7	6
16	8	5

L3={1,2,3,4,5,6...

Figure 1. 13

1.4.2 Turn Off Old Plots

Press [2nd] [STAT PLOT] ; select [4: PlotsOff] then [ENTER] . See Figures 1.14 and 1.15.

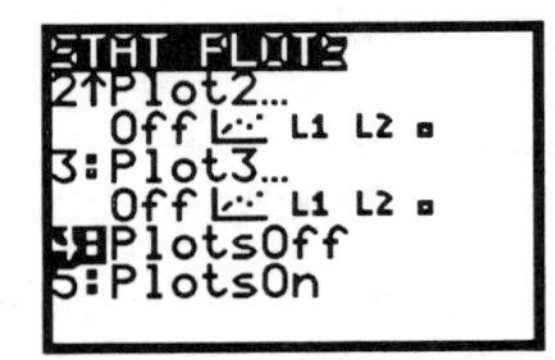

Figure 1. 14

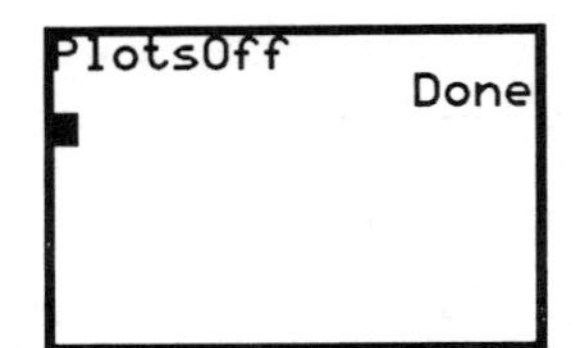

Figure 1. 15

1.4.3 Setting up Plot2

Press [2nd] [STAT PLOT] ; select [2: Plot2].

1. Select **ON** press [ENTER] .
2. Select the graph Type: **histogram icon** .
   Press [ENTER] .
3. Select the Xlist **L2** or TI-83 users [2nd] [L2] (for your horizontal axis). Press [ENTER] .
4. Select the frequency **L3**. Press [ENTER]
   In Example 2 each item has a frequency located in L3.

Set up the plot as in Figure 1.16 or 1.17.

Figure 1. 16 TI-82 Plot2 menu.

Figure 1. 17 TI-83 Plot2 menu.

1.4.4 Size the Window

Size the window to accommodate the data in Table 1.2 . Press [WINDOW] [▽] ; then enter the window settings as in Figure 1.18. Press [GRAPH] .

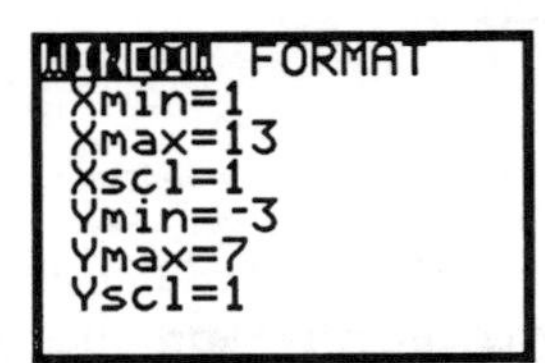

Figure 1. 18

Note: **Automatic Window Sizing.** You can have the calculator automatically set your window by pressing ZOOM ,then select [9:ZoomStat]. See Figure 1.19. You lose control of the max and min values, however, and you may still want to adjust the window.

Figure 1. 19

1.4.5 Tracing on the Histogram

Press GRAPH then TRACE and use your right arrow ▷ until your screen looks like Figure 1.20 or 1.21 . When you TRACE on a histogram, the cursor moves to the top center of each interval column. Your histogram represents the possible ways of getting each outcome by tossing a pair of dice. The P2 tells that you are tracing on Plot2. The min = 6, max = 7 indicates that you are tracing on the interval 6 to 7 or, in this case, 6. The **n** value (frequency) is 5. This means there are five possible ways to get the number six when a pair of dice is thrown.

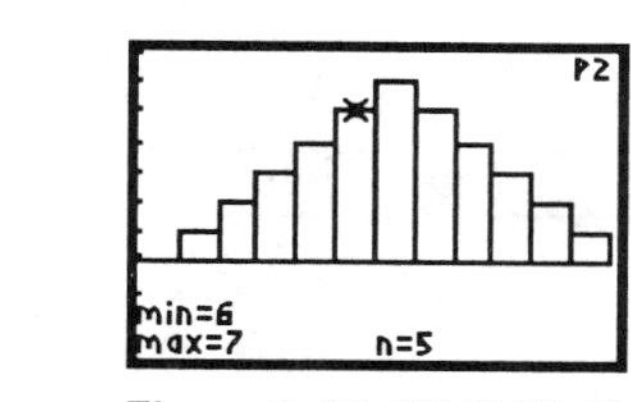

Figure 1. 20 TI-82 Plot2.

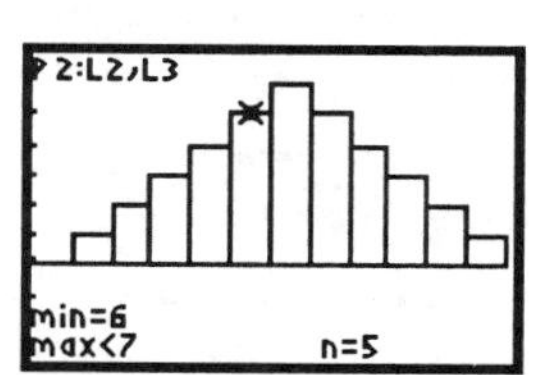

Figure 1. 21 TI-83 Plot2.

Trouble Shooting: Dimension Mismatch. The error message shown in Figure 1.22 is telling you that your lists are not the same size. Either one list is longer than the other or you have set up your plot with the wrong lists. To graph two lists they must have the same number of elements.

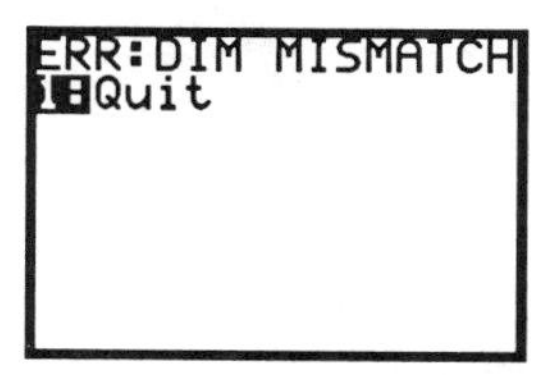

Figure 1. 22

1.5 Changing the Interval Width.

Change your WINDOW to the settings as in Figure 1.23. When you change the **Xscl**, you are setting the width size of the interval. In this case the histogram will be 2 units wide, beginning at **Xmin**.

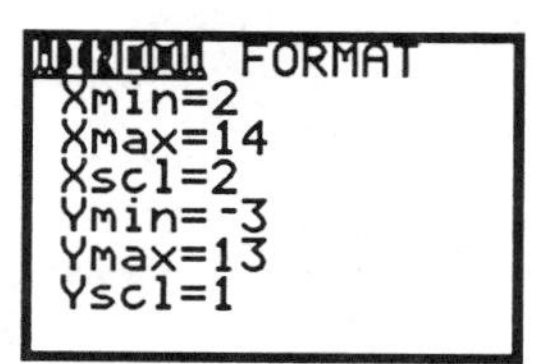

Figure 1. 23

Press GRAPH . Figure 1.24 shows the new histogram with interval width of 2 units. You would interpret the TRACE point to mean there are 11 ways to get a 6 or 7 on the roll of two dice.

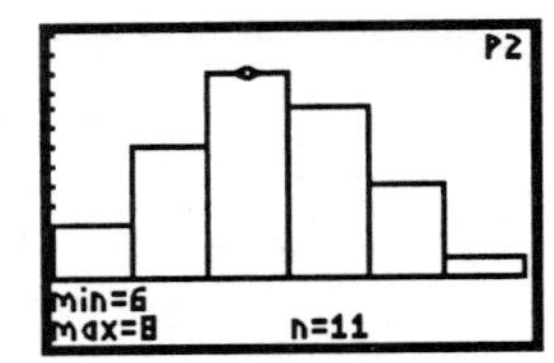

Figure 1. 24

Change your WINDOW so that **Xscl** = 3 and **Ymax** = 16.

Figure 1.25 shows the histogram with an interval width of 3 units. Press TRACE ▷

The second interval has n = 15. This means there are 15 ways to get a 5,6, or 7 on a roll of two dice.

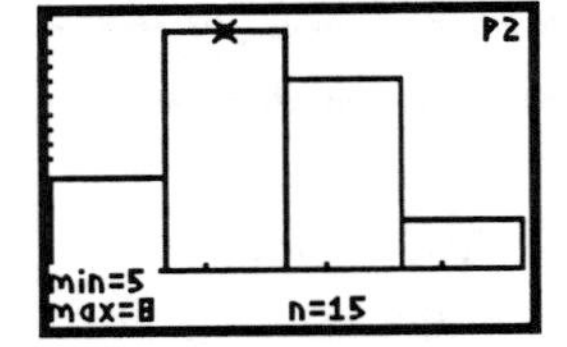

Figure 1. 25

1.6 One Variable Statistics Mean and Median

Example 3

Below are SAT scores of 10 randomly selected students.

SAT	600	640	430	500	510	530	550	370	500	530

Table 1. 3

Enter this data using the STAT menu. First **CLEAR** all values from L1, L2, and L3 using the techniques learned in Section 1.1 and 1.4 above. See Figure 1. 26.

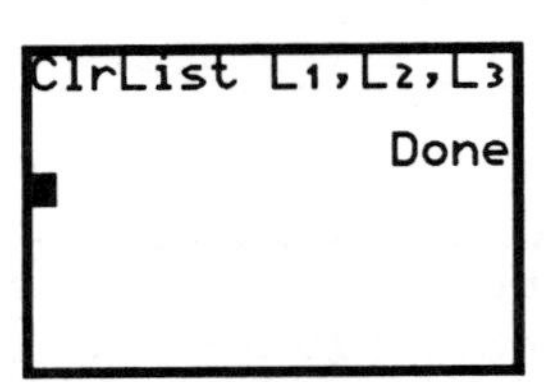

Figure 1. 26

Enter the SAT scores in L1. Refer to Section 1.1 if you need help. Press ENTER after each number. See Figure 1.27.

L1	L2	L3
600	------	------
640		
430		
500		
510		
530		
550		

L1={600,640,430...

Figure 1. 27

1.6.1 One Variable Statistics.

Let us look at the *one-variable* statistics performed on List1 (L1).

Press: STAT ▷ to < CALC>. Select [1:1-Var Stats]. See Figure 1.28. Press 2nd L1 ENTER . See Figure 1.29.

Note: To perform *one-variable* statistics on L2 or any other list, select [1:1-Var Stats] followed by the list number or name.

Figure 1. 28

We see the command line in Figure 1.29; press ENTER to see the L1 *one-variable* statistics information in Figure 1.30.

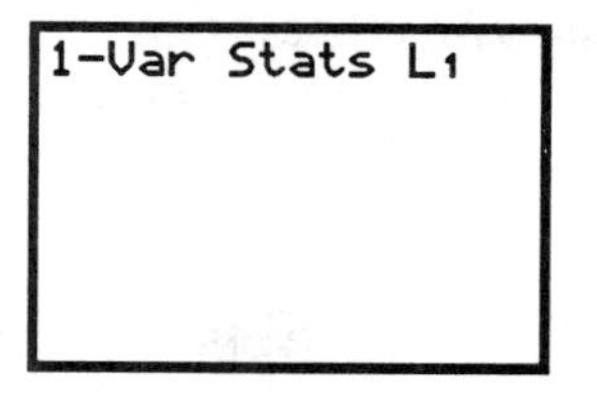

Figure 1. 29

1.6.2 The Mean.

In Figure 1.30 the statistics we are most interested in are:

1. $\bar{x}$, the **mean** (average) represented by $\bar{x}$ = **516** and
2. **n** , the number of elements in the list , **n** = 10,

This means that 516 was the average (mean) SAT score for the ten students.

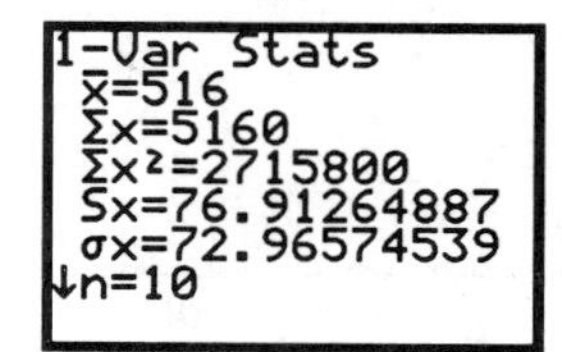

Figure 1. 30

1.6.3 The Median

Press ▽ to see the information in Figure 1.31. We are interested in **Med**, the **median** score. The median, represented by **Med = 520**, is the middle score when the data is in rank order. This means that of the ten SAT scores, five will be above 520 and five SAT scores will be below 520.

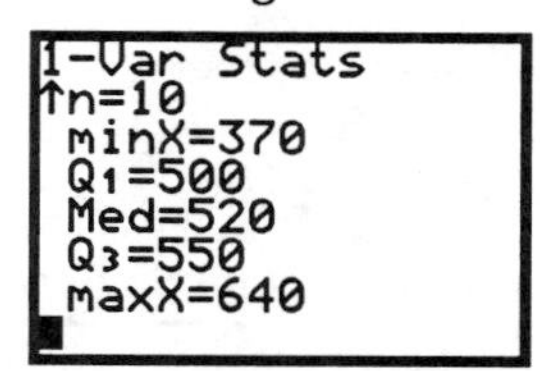

Figure 1. 31

1.6.4 Sorting a List.

Sort L1 from low to high. Press STAT , select [2:SortA(] . See Figure 1.32. Type the list you want to sort. Press 2nd L1) ENTER . See Figure 1.33. The sorted L1 is shown in Figure 1. 34.

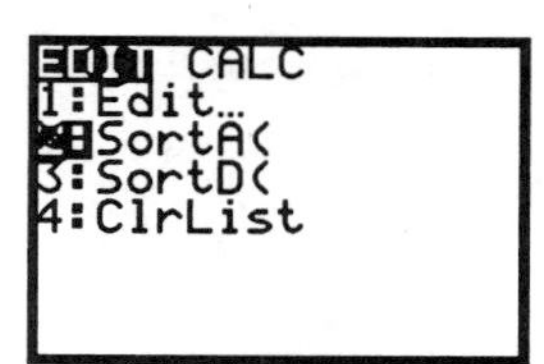

Figure 1. 32

> <u>**Trouble Shooting:**</u> If you have two lists L1 and L2 that you want to sort as an ordered pair, type **SortA(L1,L2)** ENTER . L1 will be sorted in ascending order with L2 as its paired list.

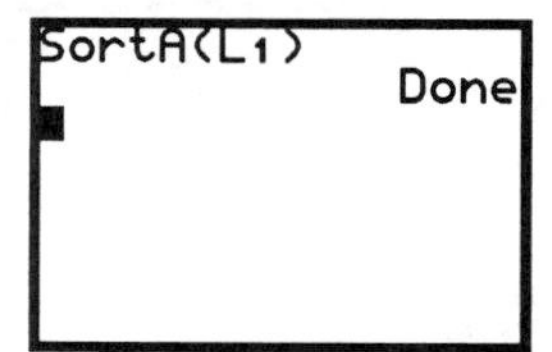

Figure 1. 33

1.6.5 Looking for the Median.

Press STAT ; select [1:Edit]. Look at your list now. See Figure 134. The **median** score is the middle score on the list, however, since there are 10 elements (an even number) the median is the score in between the fifth and sixth score, or

$$\frac{510 + 530}{2} = 520$$

Figure 1. 34

> <u>Note</u>: You may think of the median of an even number of elements as the average of the middle two scores.

1.7 Visualizing the Median; The Box and Whiskers Plot

Do a box and whisker plot. Select a plot. Turn plots OFF, press [2nd] [STAT PLOT] [4: PlotsOff] [ENTER] . Press [2nd] [STATPLOT] ; select [1:Plot1].

1. Select: **ON** [ENTER] [▽] .
2. Select graph type: [▷] to the **Boxplot** icon [ENTER] . See Figure 1.35.

Note: TI-83 icons are slightly different.

3. Select Xlist: **L1** [ENTER] .
4. Select Frequency: Choose **1** as the frequency. See Figure 1. 35.

Figure 1. 35

To graph the Boxplot: press [ZOOM] ; select [9:ZoomStat] . See Figure 1.36. The Boxplot shows one-variable statistics. Press [TRACE] The middle of the box is the median (**Med**). See Figure 1.37. The whiskers on the plot extend from the minimum list value on the left to the first quartile (**Q_1**) and from the third quartile (**Q_3**) to the maximum list value. Use [◁] and [▷] arrows to view these scores. See Figure 1.31 for the Q_1, Med, and Q_3 numerical display).

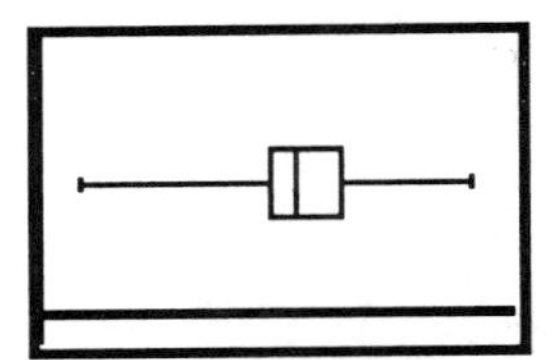
Figure 1. 36

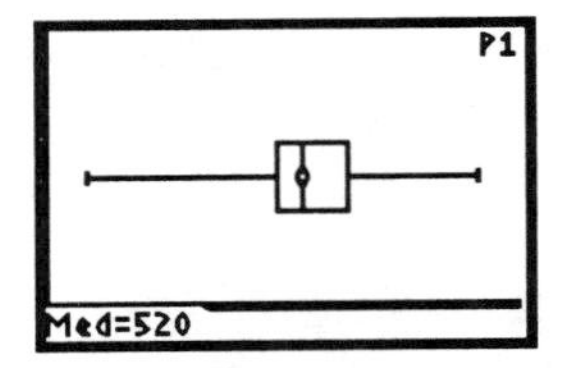

Figure 1. 37

Note: ***Quartiles*** are 1/4 or 25% of the list when put in rank order.

1.7.1 Show a Histogram of L1.

Make a histogram. Select the plot. Press [2nd] [STATPLOT] ; select [1:Plot 1].

1. Select: **ON** [ENTER] .
2. Select: **(histogram icon)** [ENTER] .
3. Xlist: **L1** [ENTER] .
4. Freq: Select **1** . See Figure 1.38.

Figure 1. 38

Press [WINDOW] [▽] . Select an appropriate range and interval scale based on the data in L1. See Figure 1.39.

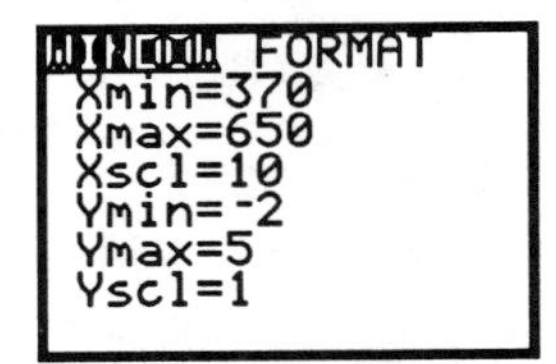

Figure 1. 39

Press [GRAPH] . Press [TRACE] to view the frequency, **n**, of each SAT score. See Figure 1. 40.

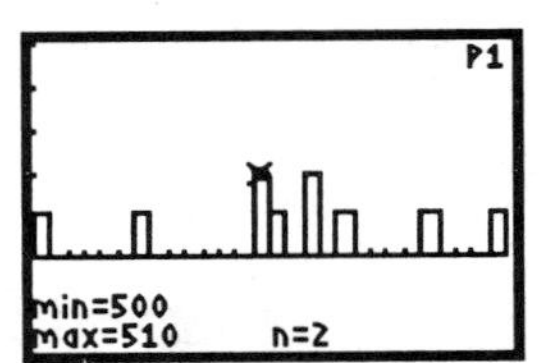

Figure 1. 40

1.8 Other List Techniques
Example 4

Find the mean age of students in a mathematics class.

Age Interval	Frequency Count
15 - 19	2
20 - 24	8
25 - 29	4
30 - 34	3
35 - 39	2
40 - 44	1
45 - 49	1
Total	21

Table 1. 4

Table 1.4 gives the age of students. Since the age is given as an interval, <u>calculate the median of each interval</u> so that a single number can be entered into a list (i.e. 17 is the median of 15 - 19). Go to the home screen and type the median data from Table 1.4 in L1. Remember to enclose the list in braces, { } as in Figure 1.41.

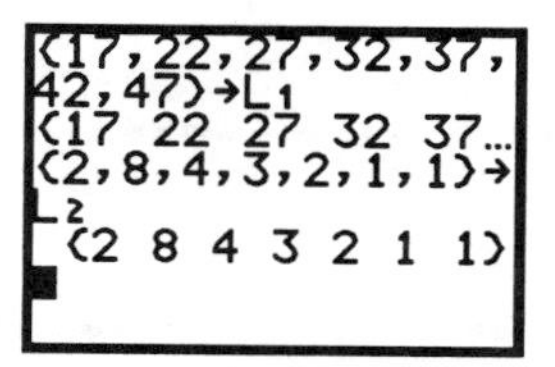

Figure 1. 41

Press [2nd] [{] 17, 22, 27, 32, 37, 42, 47 [2nd] [}] [STO] [2nd] [L1] [ENTER] . Store the frequency in L2. Press [2nd] [{] 2, 8, 4, 3, 2, 1, 1 [2nd] [}] [STO] [2nd] [L2] [ENTER] . See Figure 1.41.

> **Note**: To calculate the mean you need to multiply the age times the frequency, sum all the ages then divide by the total frequency. See Figures 1.42 through 1.45.

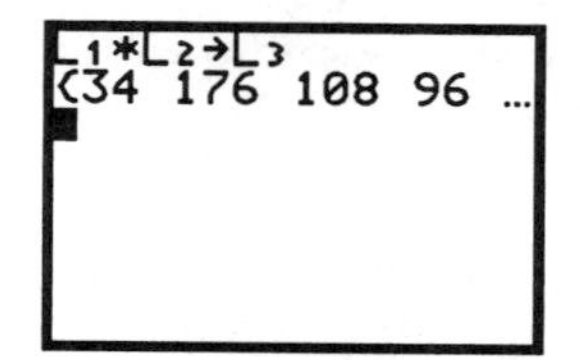

Figure 1. 42

Calculate the *mean* from the Home Screen.

1. **Multiply L1 x L2; then store to L3.**

[2nd] [L1] [X] [2nd] [L2] [STO] [2nd] [L3]

See Figure 1.42.

2. **Sum L3.**

[2nd] [LIST] [▷] to <MATH>; select [5:sum]

[2nd] [L3] [ENTER] . See Figures 1.43 and 1. 44.

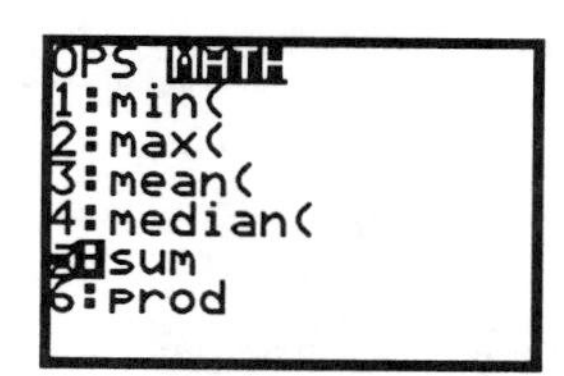

Figure 1. 43

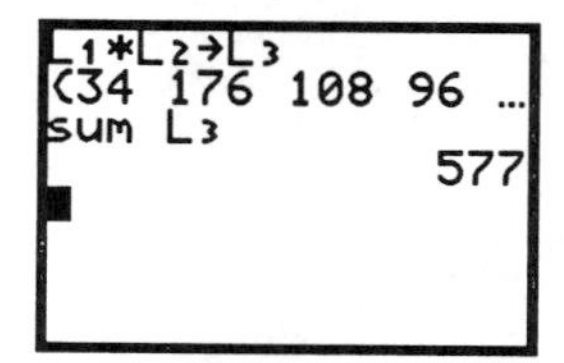

Figure 1. 44

3. **Divide by the Total Number .**

Press [÷] 21 . See Figure 1. 45.

1.8.1 Use the Mean Command for Two Lists

[2nd] [LIST] [▷] to <MATH>, select [3:mean],

[2nd] [L1] [,] [2nd] [L2] . See Figure 1.45.

Either method gives the same answer.

```
{34 176 108 96 …
sum L3
                577
Ans/21
        27.47619048
mean(L1,L2)
        27.47619048
```

Figure 1. 45

1.9 Scatter Plots

Relationships between two variables can be visualized by graphing data as a scatter plot. Think of the two list as ordered pairs. An ordered pair (x, y) can represent a point on a graph.

Example 5

In Table 1.5 are the SAT and math placement scores of ten randomly selected freshmen students. Graph the data to see if a relationship exits between the SAT scores and the math placement scores.

SAT	600	640	430	500	510	530	550	370	500	530
Math Place	25	29	14	12	11	8	17	16	9	26

Table 1.5

Note: Refer back to Chapter 1 Sections 1.1 through 1.4 for information on entering data into lists.

1.9.1 Enter the Data

CLEAR List 1 (L1)and List 2 (L2).
Enter the math placement scores into L1.
Enter the SAT scores into L2 . See Figure 1.46.

L1	L2	L3
25	600	------
29	640	
14	430	
12	500	
11	510	
8	530	
17	550	

L2={600,640,430...

Figure 1.46

1.9.2 Set Up the Scatter Plot

Press [2nd] [STATPLOT] . Select [1:Plot1].

Set up the Scatter Plot as in Figures 1.47 or 1.48.

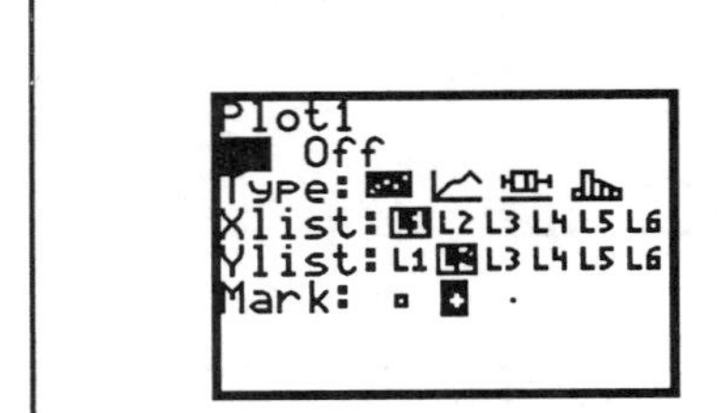

Figure 1.47

TI-82 Plot1 setup.

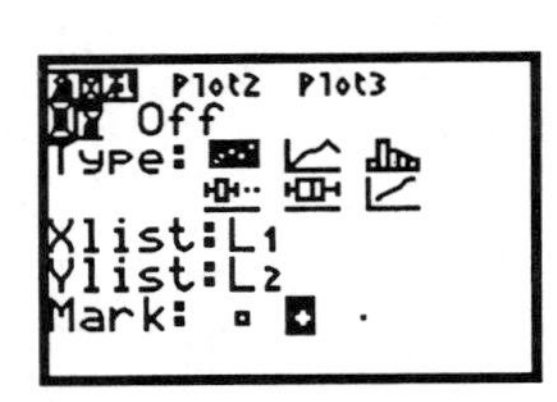

Figure 1.48

TI-83 Plot1 setup.

1.9.3 Size the Window

Press [WINDOW] [▽] . Set the appropriate window based upon the data in L1 and L2, as in Figure 1.49.

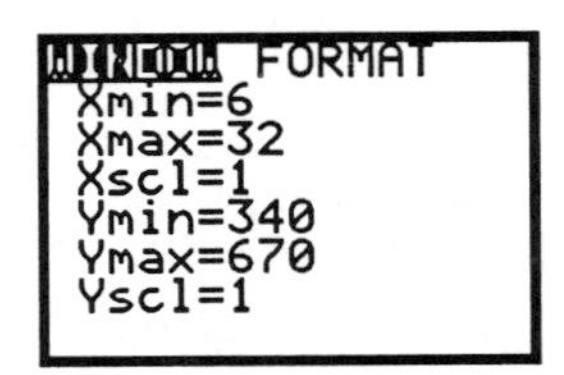

Figure 1.49

The **X list** contains the math placement scores and the **Ylist** contains the SAT scores, so the window is set just beyond the lowest and highest data values.
X: [Xmin, Xmax] = X: [6, 32] and
Y: [Ymin,Ymax] = Y: [340, 670].

To view the Scatter Plot press GRAPH .

See Figure 1.50. There appears to be an increasing relationship, i.e. as the math placement score increases the SAT score increases, but there are a few exceptions (outliners).

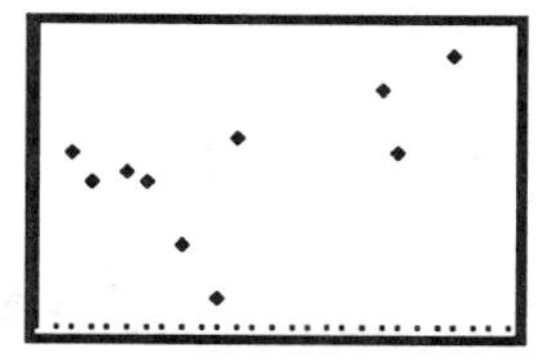

Figure 1. 50

Math placement on the x-axis.

1.9.4 Switch the x and y Variables.
What does the plot look like when L2, the SAT score, is the independent variable? Set up the plot as in Figure 1.51 or 1.52.

Plot1
On Off
Type:
Xlist:L1 L2 L3 L4 L5 L6
Ylist:L1 L2 L3 L4 L5 L6
Mark:

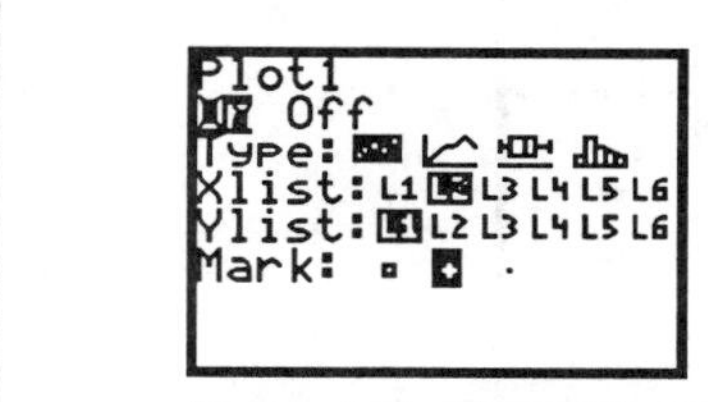

Figure 1.51 TI-82 Plot1 setup.

Plot1 Plot2 Plot3
On Off
Type:
Xlist:L2
Ylist:L1
Mark:

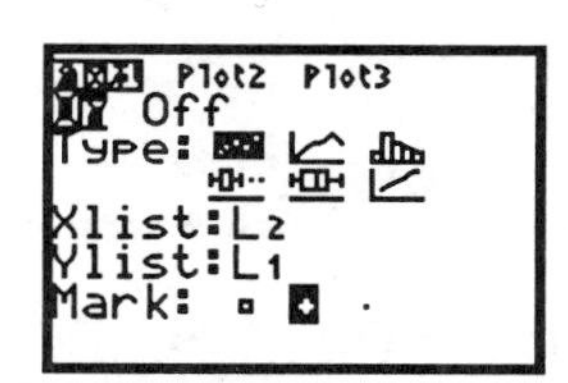

Figure 1.52 TI-83 Plot1 setup

1.9.4 Sort List 2 and List 1 Together
To sort List2 while keeping the pairing with List1, Press STAT , select [2:SortA]

2nd L2 , 2nd L1) ENTER .

See Figure 1.53.

SortA(L2,L1)
Done

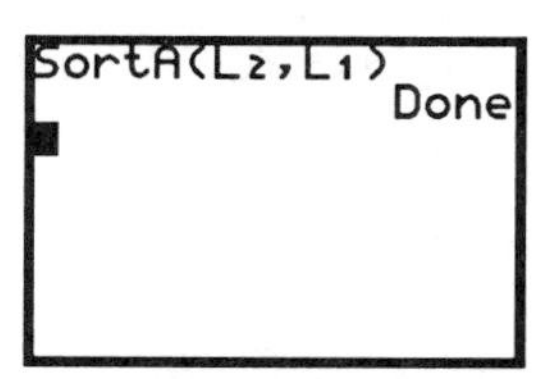

Figure 1.53

1.9.4 Setting the Window
Look at the sorted list. See Figure 1.54.

Press STAT ; select [1:Edit]. Use the sorted List2 to determine the window setting . Remember we want L2 to be the independent variable. The window is set just beyond the lowest and highest data values. X: [350, 660] and Y: [0, 35].

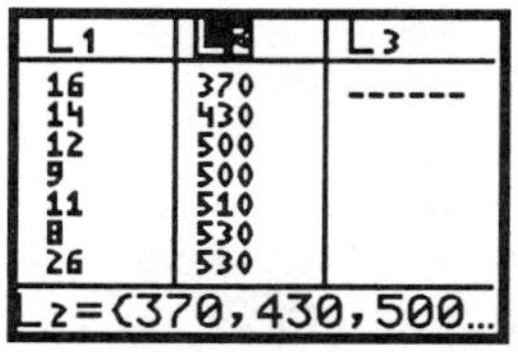

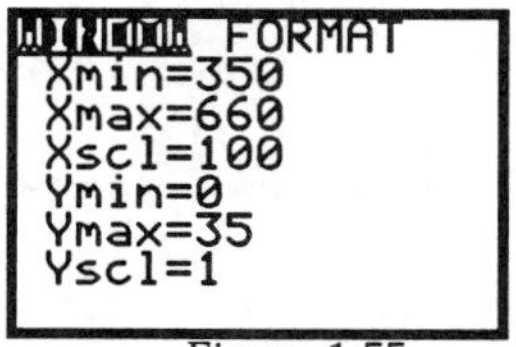

Press WINDOW ∇ . Set the window as in Figure 1.55.

Note: There are many correct window sizes. Choose a window that will contain all of the points and slightly beyond the points.

L1
L2
L3
16 370
14 430
12 500
9 500
11 510
8 530
26 530

L2={370,430,500...

Figure 1.54

WINDOW FORMAT
Xmin=350
Xmax=660
Xscl=100
Ymin=0
Ymax=35
Yscl=1

Figure 1.55

Trouble Shooting: Before you press graph:

1. Clear [Y=] or turn **OFF** all graphs. To turn off graphs, place the cursor on the = sign then press [ENTER] .
2. Turn **OFF** all plots except the one you want to see.

ERR: INVALID DIM means your lists are not the same size, you have selected a list with no data in it or you have a plot turned on that you did not want and it has different size lists.

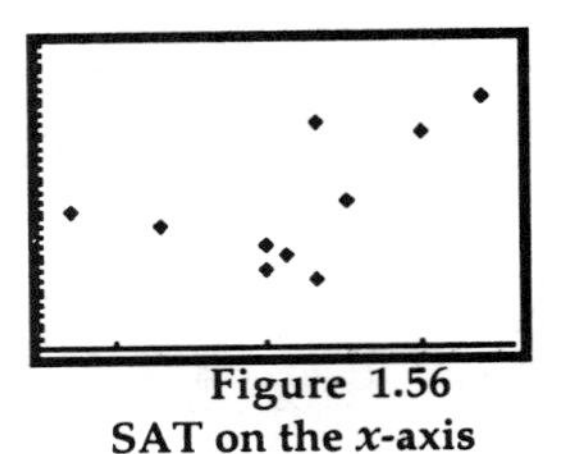

Figure 1.56
SAT on the x-axis

Press [GRAPH] . See Figure 1.56. Here the relationship is less clear. Perhaps more data is necessary to determine a trend or relationship. Explore the plot using **TRACE** . See Figure 1.57.

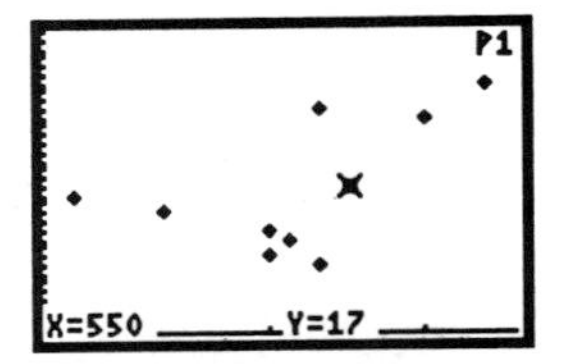

Figure 1.57

1.9.6 A Check List for Plotting

To plot statistical data in lists, follow these steps:

1. **Clear old data in lists.**
2. **Store the statistical data in one or more lists.**
3. **Set up the STAT PLOT.**
4. **Turn Plots ON or OFF as appropriate. See Figure 1.58.**
5. **Clear or deselect [Y=] equations as appropriate . See Figure 1.59.**
6. **Define the viewing WINDOW.**
7. **Explore the plot or graph by pressing TRACE . See Figure 1.57.**

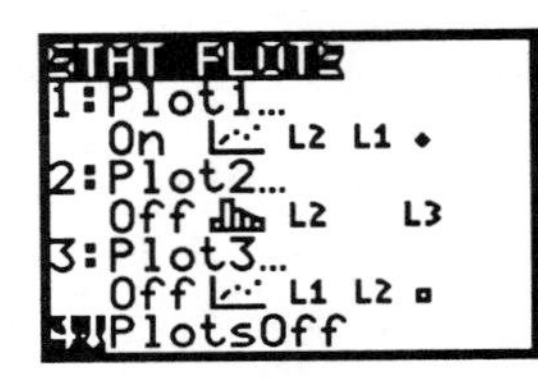

Figure 1.58

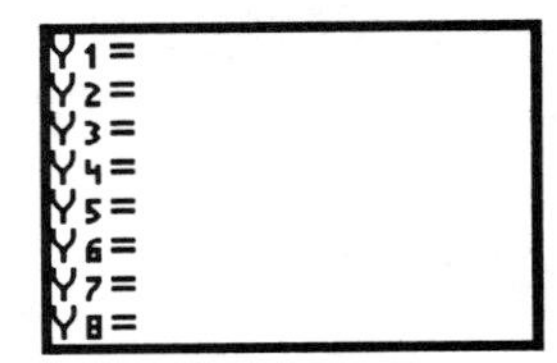

Figure 1.59

1.10 Introduction to Graphing Functions

The graphing calculator can be used to graph equations that are functions. The top row of keys, under the viewing screen , contains all the graphing menus.

Note: Before you begin graphing

1. Turn all plots **OFF**: [2nd] [STATPLOT] select [4 :PlotsOff] [ENTER] . See Figure 1.58. The calculator will say ***Done***. .
2. Press [Y=] and [CLEAR] all equations. See Figure 1.59.

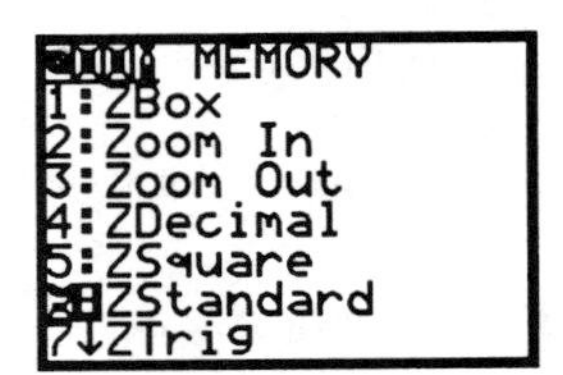

Figure 1.60

1.10.1 The Standard WINDOW

Press [ZOOM] select [6:Zstandard]. See Figure 1.60). The graph screen and the x y coordinate system appears. See Figure 1.61.

You see only a portion of the real number line on the xy-coordinate plane. The size of the viewing window is determined by the window variables : **Xmin, Xmax, Ymin,** and **Ymax** . See Figure 1.62).

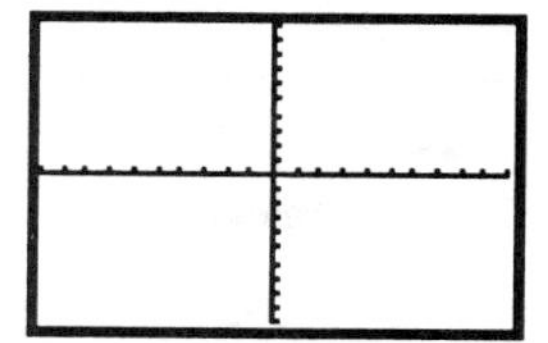

Figure 1.61

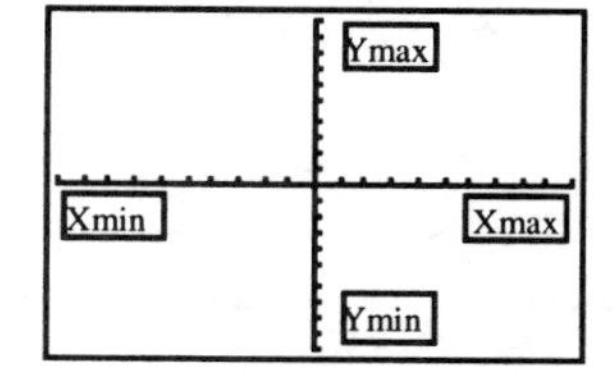

Figure 1.62

(The words will not appear on your screen)

To see the current ***Standard Window*** , press WINDOW . See Figure 1.63).

The distance along the X-axis goes from -10 to 10 or X: [-10, 10]. The distance along the Y-axis goes from -10 to 10 or Y: [-10, 10]. The distance between the tic marks on the *x*-axis is 1 unit (Xscl = 1) and the distance between the tic marks on the *y*-axis is 1 unit (Yscl = 1).

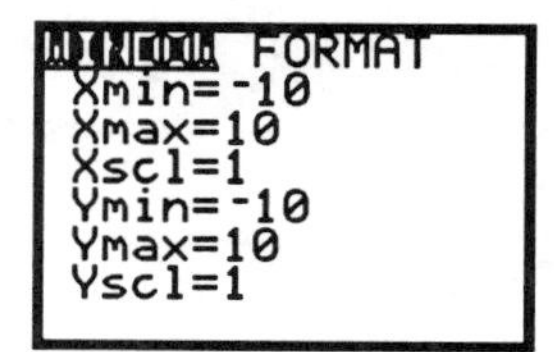

Figure 1.63

1.10.2 The Free Moving Cursor

Press GRAPH . When you press the arrow keys, ▷ ◁ △ ▽ , the cursor can move anywhere on the graphing window. The cursor has changed to a cross and at the bottom of the screen you see the coordinates of the screen position, which change as you jump form pixel to pixel. A pixel is a point of light on the screen. See Figure 1.64.

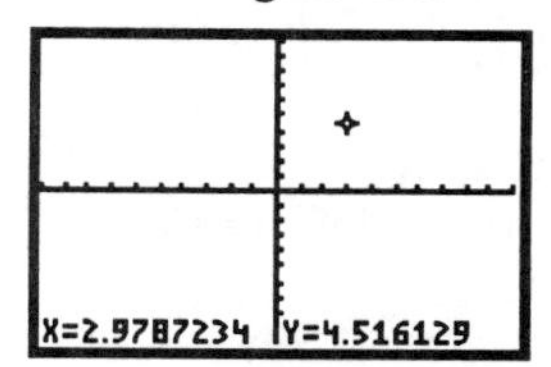

Figure 1.64

1.10.3 The Decimal Window

Notice that on the standard window you get rather "ugly" decimals when you press the arrow keys.

Press ZOOM ; select [4:ZDecimal].

Now press the arrow keys. As you jump from pixel to pixel you increment by the decimal value 0.1 (1/10). See Figure 1.65.

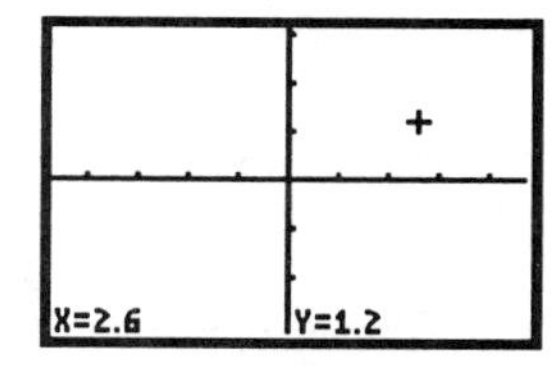

Figure 1.65

Press WINDOW to see the decimal window settings for *x* and *y* . See Figure 1.66.

WINDOW FORMAT
Xmin=-4.7
Xmax=4.7
Xscl=1
Ymin=-3.1
Ymax=3.1
Yscl=1

Figure 1.66

Example 6

To change Centigrade temperature to Fahrenheit use the formula F = (9/5)C +32. Enter this formula into the calculator and graph the function.

Press Y= ; into **Y1** type **(9/5)** *x* **+ 32.**

See Figure 1.67 or 1.68.

Trouble Shooting: The independent variable, in this case C, must be entered as X on the calculator. The dependent variable F will become Y. Notice you can graph 10 different functions, Y1 to Y10.

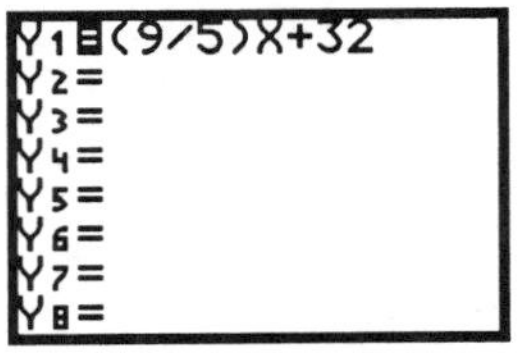

Y1=(9/5)X+32
Y2=
Y3=
Y4=
Y5=
Y6=
Y7=
Y8=

Figure 1.67

The TI-82 [Y=] Screen

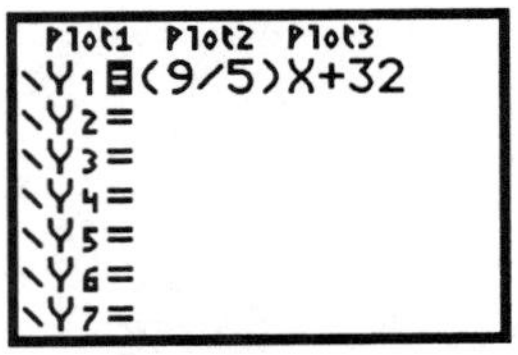

Plot1 Plot2 Plot3
\Y1=(9/5)X+32
\Y2=
\Y3=
\Y4=
\Y5=
\Y6=
\Y7=

Figure 1.68

The TI-83 [Y=] Screen

A Special Note to TI-83 Users.

Plots can be turned **ON** and **OFF** from the [Y=] screen. To turn a plot ON, use the arrow keys to position the cursor on the desired plot name (Plot1, Plot2, Plot3), then press [ENTER]. To turn the plot OFF press [ENTER]. Darkened plots are ON.

1.10.4 The Integer Window

Press [WINDOW] [∇], then set the window to the settings as in Figure 2.24.

Press [GRAPH], then [TRACE]. Right arrow to 8° C to see the equivalent temperature 46.4° F. See Figure 1.70. Left arrow to -10° C for the equivalent 14° F. See Figure 1.71. Notice the pixel jumps are now integers of two units. Play with the arrow keys as you TRACE on the function.

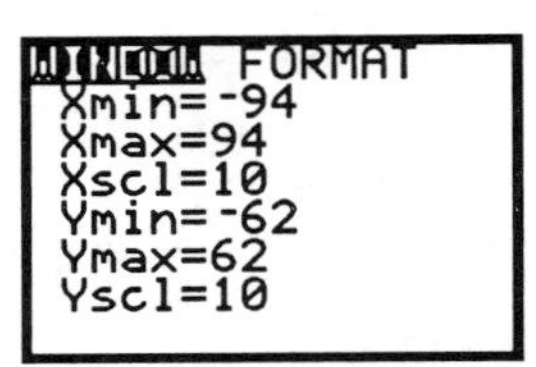

Figure 1.69

Friendly Windows

As you move in an x direction from pixel to pixel there are 94 jumps across the screen. Likewise there are 62 pixel jumps in a y direction. To determine the horizontal jump, Δx, and the vertical jump, Δy, use the following formulas:

$$\Delta x = \frac{x_{max} - x_{min}}{94}$$

$$\Delta y = \frac{y_{max} - y_{min}}{62}$$

Note: A "Friendly Window" is any window whose distance between **Xmax** and **Xmin** is evenly divisible by 94. In the above window $\Delta x = (94 - (-94)) / 94 = 2$, or two units per each x pixel jump.

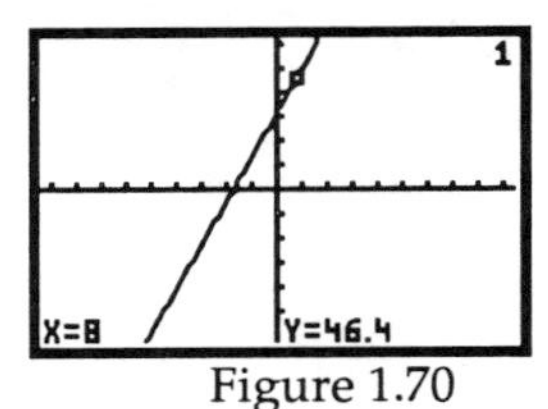

Figure 1.70

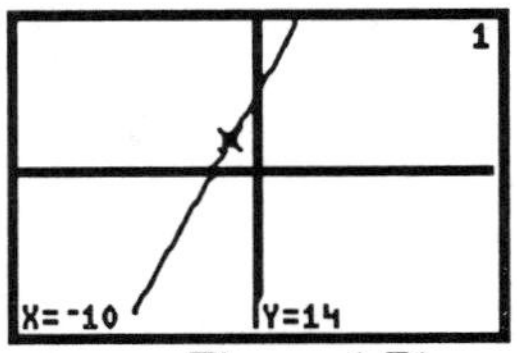

1
X=-10
Y=14

Figure 1.71

1.10.5 Use Table to Help Find the Appropriate Window.

Finding the appropriate window for a function takes a lot of practice. The following tips can be very helpful:

1. Use algebra to calculate a few easy values of y, such as when $x = 0, 1, -1$ etc.
2. Press [TRACE] to discover where a few points lie on the graph.
3. Use [2nd] [TABLE] to see many values of x and y.

Example 7

Find an appropriate window for
$f(x) = y = x^2 + 25$

Press [Y=] ; into **Y1** type: $x^2 + 25$. See Figure 1.72. Press [ZOOM] ; select [6:ZStandard]. No graph is seen. See Figures 1.73 and 1.74.

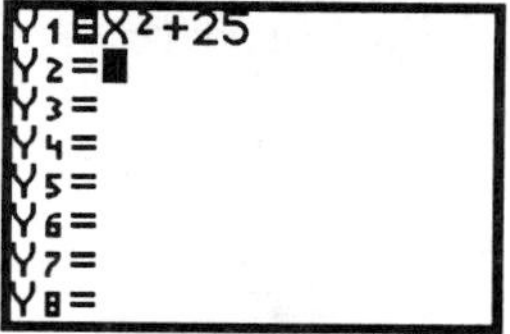

Figure 1.72

Figure 1.73

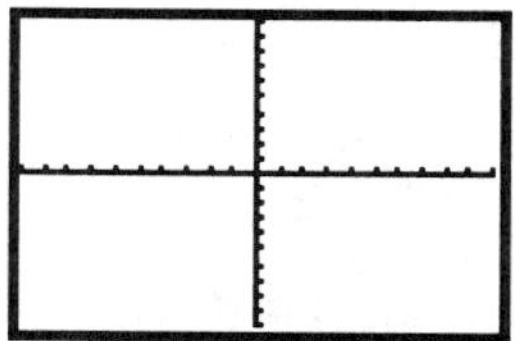
Figure 1.74

Algebra tells us that when $x = 0$, $y = 25$. On the standard viewing window the y values are Y: [-10, 10], so $y = 25$ cannot be seen!

Press [TRACE] ; hold down [▷] to confirm that y is larger than 25. See Figure 1.75. Press [2nd] [TblSet] . Begin the table at $x = -5$ and increment by 1 unit ($\Delta tbl = 1$). Set up a table of values as in Figure 1.76 . Press [2nd] [TABLE] . Use [▽] to scroll through some values. See Figure 1.77.

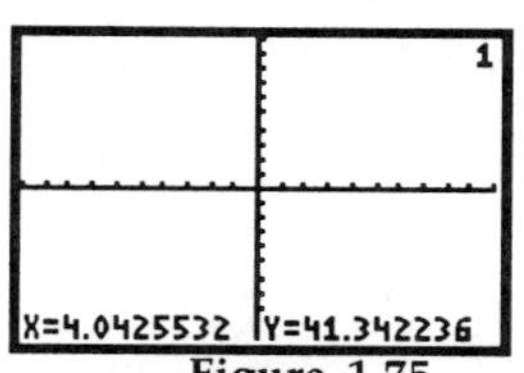

Figure 1.75
TRACE reveals points on the graph.

Figure 1.76

X	Y1
-5	50
-4	41
-3	34
-2	29
-1	26
0	25
1	26

X=1

Figure 1.77

An appropriate window based upon the table appears to be X :[-10, 10] and Y: [-10, 50] or better still Y: [-10, 100], with Yscl = 10. See Figure 1.78.

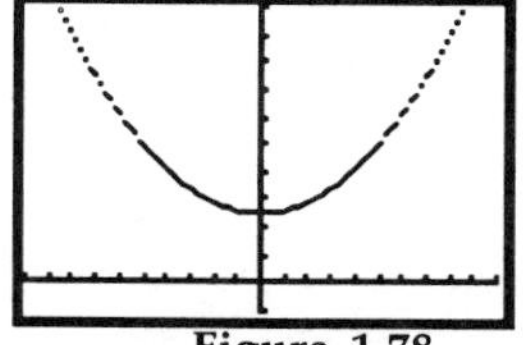
Figure 1.78
X: [-10, 10] and Y: [-10, 100]

Trouble Shooting: To lessen the trial and error process for finding an appropriate viewing window, always use **TRACE** and **TABLE** to find points on your graph. Usually it is the Ymax that needs to be adjusted not Xmax.

1.11 Using Graphs To Determine the Domain of a Function

For most functions the ***domain*** (possible x values) is all real numbers. However there are functions that are the exception.

Example 8

Use a graph and a table of values to help determine the domain of the function
f(x) = y = 1/x.

Press [Y=] [CLEAR]. Into Y1 type 1/x, press 1 [÷] [X,T,Θ]. See Figure 1.79. Press [ZOOM]; select [4:ZDecimal] [TRACE].

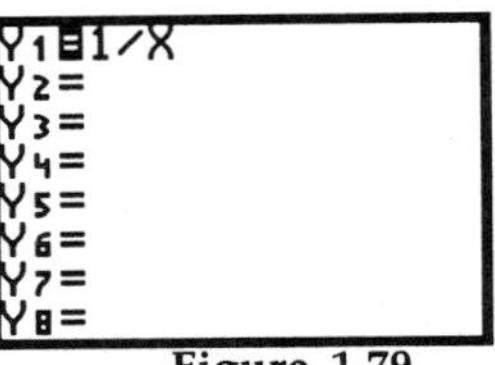

Figure 1.79

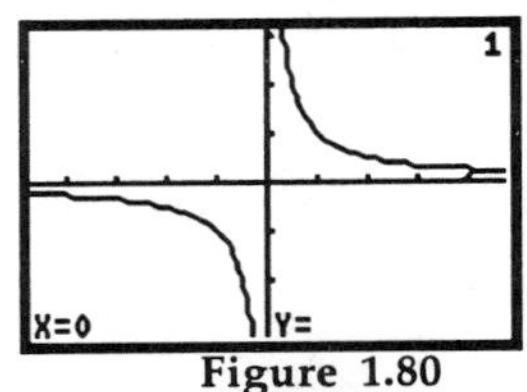

Figure 1.80
y is undefined for *x* = 0

We see a graph in two pieces. TRACE is telling us that there is no y value associated with x = 0. See Figure 1.80.

Set up a table of values. Press [2nd] [TblSet]. Begin the table at x = -5 and increment by 1 unit (Δtbl=1). See Figure 1.81. Press [2nd] [TABLE]. Notice that for x = 0 we get an ERROR message, which means that y is undefined for x = 0. See Figure 1.82.

Figure 1.81

X	Y1
-5	-.2
-4	-.25
-3	-.3333
-2	-.5
-1	-1
0	ERROR
1	1

X=0

Figure 1.82

Both the graph and the table confirm that x = 0 is not in the domain of x. The domain of $f(x) = 1/x$ is the set of all real numbers x,, $x \neq 0$. Using interval notation:

$$(-\infty, 0) \cup (0, +\infty)$$

Example 9

Use a graph and a table of values to help determine the domain of

$$f(x) = \sqrt{x}$$

Press [Y=] [CLEAR]. Into Y1 type [2nd] [√] [X,T,Θ]. See Figure 1.83.

Press [GRAPH] [TRACE].

Figure 1.83

The function $f(x)$ appears to be defined for x = 0, but not to the left of zero, $x < 0$. See Figures 1.84 and 1.85.

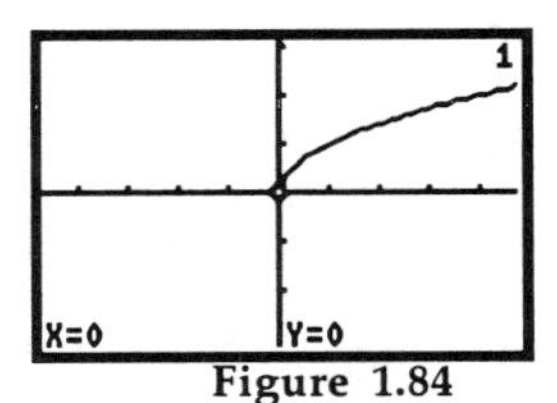

Figure 1.84

The domain of $f(x) = \sqrt{x}$ s the set of all real numbers $x \geq 0$ or x: $[0,+\infty)$. Confirm with a table of values.

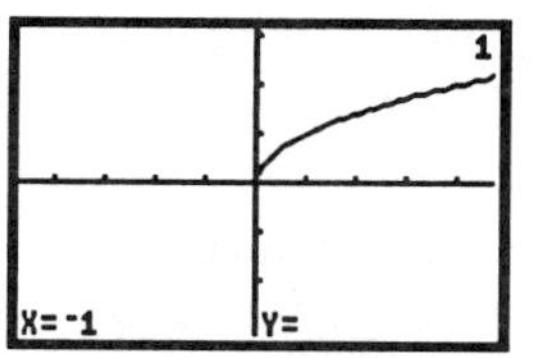

Figure 1.85

Press [2nd] [TABLE] . Notice that values for x less than 0 give an ERROR message, which means the function is undefined for $x < 0$.

See Figure 1.86. Use [▽] to see other values of y that are defined by x.

X	Y1
-5	ERROR
-4	ERROR
-3	ERROR
-2	ERROR
-1	ERROR
0	0
1	1

X= -1

Figure 1.86

This helps confirms that the domain of the square root function is the set of all x such that $x \geq 0$, i.e., domain = $\{x \mid x \geq 0\}$.

1.11.1 Graphs that are NOT Functions.

Example 10

Graph $y^2 = x$

First solve for y. There are two values for y, $y = +\sqrt{x}$ or $y = -\sqrt{x}$, therefore y is not a function. However, a relationship does exist between x and y. We need to trick the calculator into graphing the relationship.

Let **Y1** = $\sqrt{x}$ and **Y2** = -$\sqrt{x}$. . See Figure 1.87.

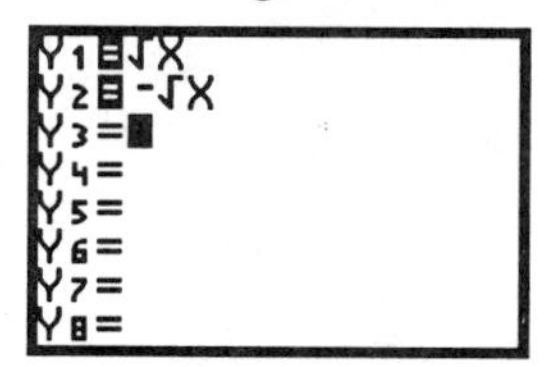

Figure 1.87

Press [GRAPH] . See Figure 1.88.

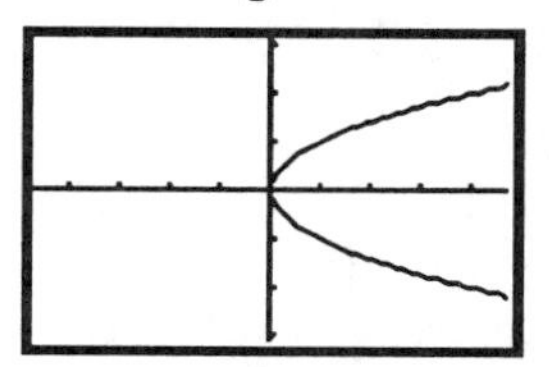

Figure 1.88

In addition we can tell y is **not** a function of x because it fails the *vertical line test*.

> **NOTE**: The **Vertical line test** says if a vertical line intersects a graph at more than one point, the graph is not a function.

Press [2nd] [DRAW] ; select [4:Vertical]
See Figure 1.89.

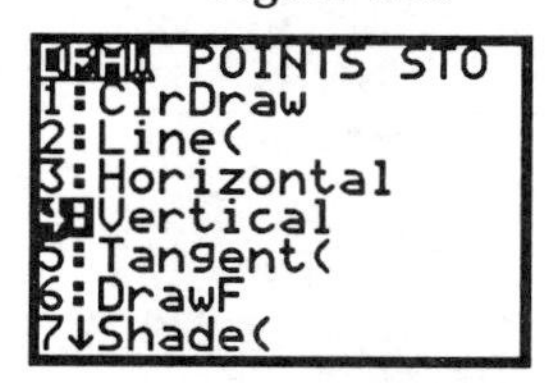

Figure 1.89

Use the right arrow [▷] to position the vertical line on $x = 3$.

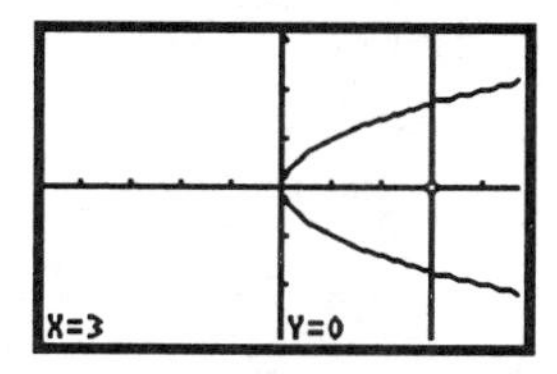

Figure 1.90

(3, 0) are the coordinates of the x-intercept of the vertical line

Since a vertical line intersects the graph in two places (i.e. there are two y values for one x value), the graph is **not** a function.
See Figure 1.90.

1.11.2 Range and Graphs of Other Interesting Functions.

Example 11

a) Graph the absolute value function

$f(x) = y = |x|$

Press [Y=] and [CLEAR] all values from Y1.

*Type [2nd] [ABS] [(] [X,T,Θ] [)] . See Fig 1.91.

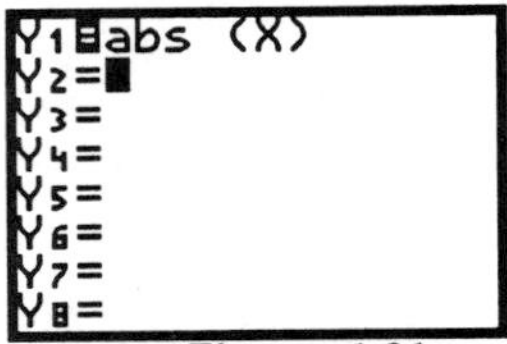

Figure 1.91

> ***TI-83 Note:** The absolute value command is under [MATH] <NUM> [1:abs( ]. The left parentheses is provided.

To graph press [ZOOM] [4:ZDecimal].

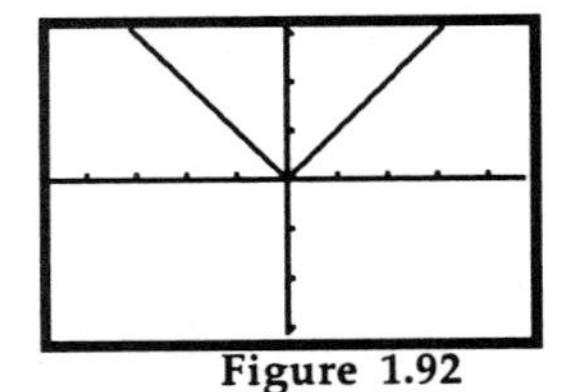
Figure 1.92

The absolute value graph has the distinctive "V" shape. The absolute value of a number is a nonnegative value. Therefore, your output values (y values), or the *range* of the function, is $y \geq 0$.

b) Graph the absolute value function

$g(x) = y = |x + 2| - 1$

Press [Y=] into Y2 *type [2nd] [ABS] [(] X+2 [)] -1 . (*TI-83 see above box). See Fig. 1.93.

Figure 1.93

Press [GRAPH] [TRACE] [▽] . You are on Y2 . Left arrow to $x = -2$. See Figure 1.94. This reveals the lowest y value on the graph, or $y = -1$. The range for $y = |x + 2| - 1$ is:

$y \geq -1$ or $\{y \mid y: [-1, +\infty)\}$

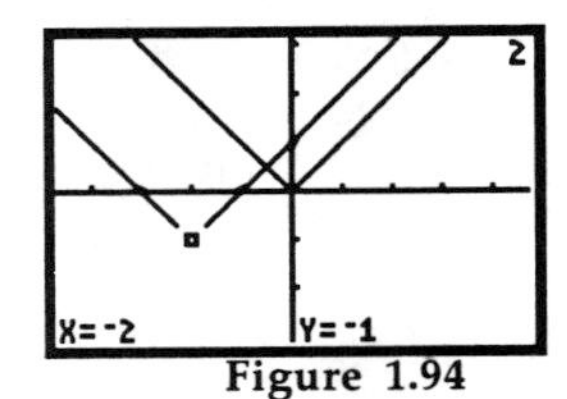

Figure 1.94

c) Graph the greatest integer function :

$h(x) = y = [[x]]$

The **greatest integer function** denoted by $[[x]]$, and is defined as the greatest integer less than or equal to x. $[[4.3]] = 4$ and $[[-2.7]] = -3$. This is the integer to the *left* of the number. The graph is often referred to as a step function. To graph press [MODE] reset the calculator to **dot mode** . See Figure 1.95.

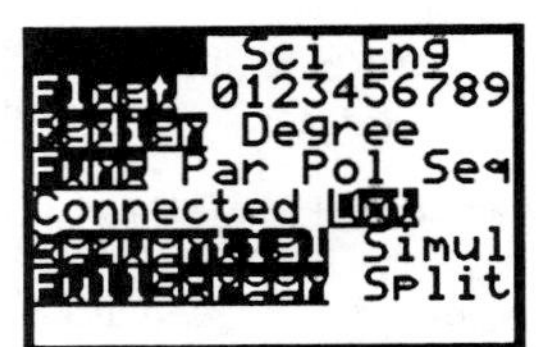

Figure 1.95

Press [Y=] and [CLEAR] all expressions.

Press [MATH] <NUM> select [: int]. Your function should be typed as in Figure 1.96.

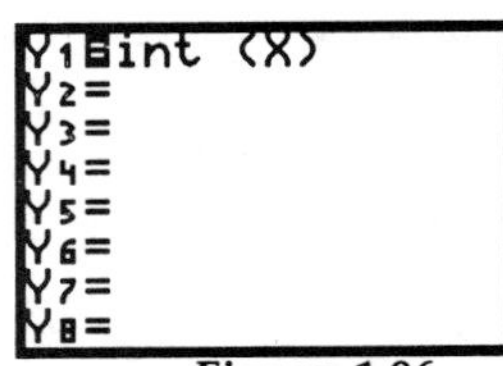

Figure 1.96

Press [GRAPH] . The distinctive step function appears. The domain for x is any real number. [TRACE] reveals that the range of y is always an integer. See Figure 1.97.

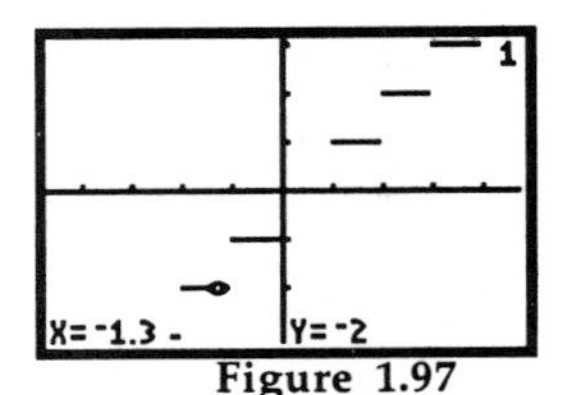

Figure 1.97

The graph shown in Dot Mode

> **Note: Remember to change back to connected MODE.**

Chapter 2
Average Rate of Change and Linear Functions

2.1 Finding the Average Rate of Change

Example 1

In 1980 the US Federal debt was 909 billion dollars. In 1990 the Federal debt was 3206 billion dollars. Find the average rate of change.

The average rate of change is:

$$\frac{\text{change in debt}}{\text{change in years}} = \frac{3206 - 909}{1990 - 1980} = 229.7$$

or 229.7 billion dollars per year.
This means on average, the US Federal debt increased by $229.7 billion/yr. from 1980 to 1990. See Figure 2.1.

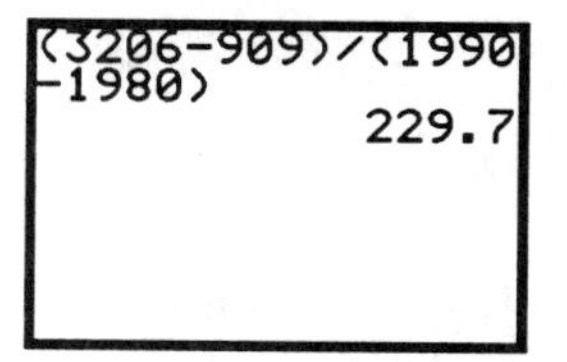

Figure 2. 1

2.1.1 Using Lists to Find the Rate of Change

Enter the data from Example 1 in list 1, L1, and list 2, L2, then calculate the average rate of change.

1. Press [STAT] ; select [4:ClrList];
2. Type: L1, L2, L3 [ENTER] . See Figure 2.2.
3. Enter the data in L1 and L2. Press [STAT] ; select [1:Edit]. Remember to press [ENTER] or [▽] after every entry . See Figure 2.3.

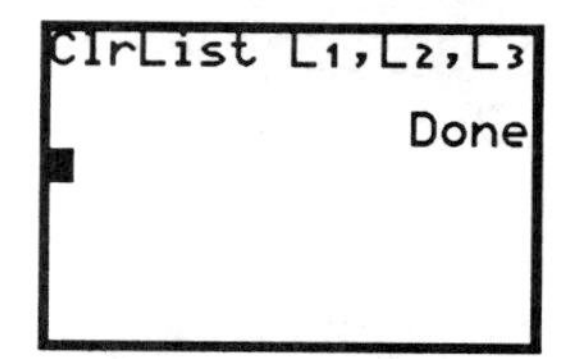

Figure 2. 2

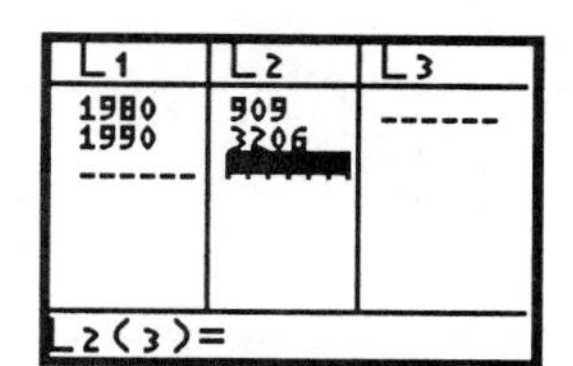

Figure 2. 3

Use the list position to calculate the average rate of change.
Type the following list equation:

$$\frac{(L2(2) - L2(1))}{(L1(2) - L1(1))}$$

Press [2nd] [QUIT] ; then type:

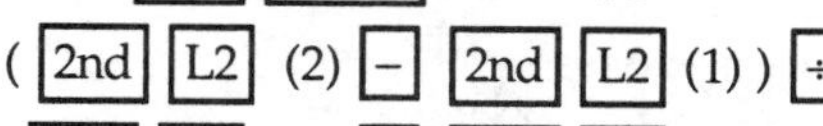

([2nd] [L2] (2) [−] [2nd] [L2] (1)) [÷]
([2nd] [L1] (2) [−] [2nd] [L1] (1)).

See Figure 2.4. While this may seem tiresome to type, the advantage is that you can change the numbers in the list and then recall the rate of change list equation.

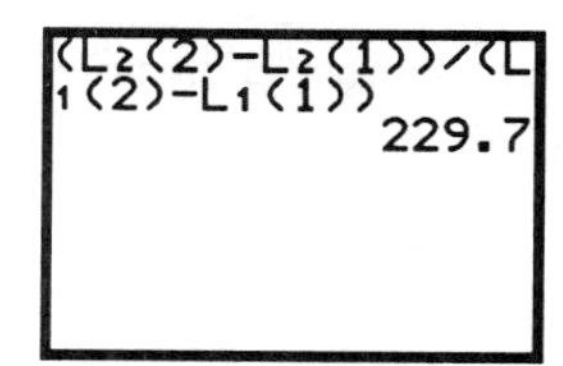

Figure 2. 4
Carefully enter all parentheses.

Example 2

In 1985 the federal debt was 1817 billion dollars. Find the average rate of change from 1985 to 1990.

a) Change L1 and L2 to look like Figure 2.5.

b) Recall the rate of change equation on the Home Screen.

Press [2nd] [QUIT] [2nd] [ENTRY] [ENTER] . .

See Figure 2.6.

This means that from 1985 to 1990 the Federal debt increased on average $277.8 billion/yr.

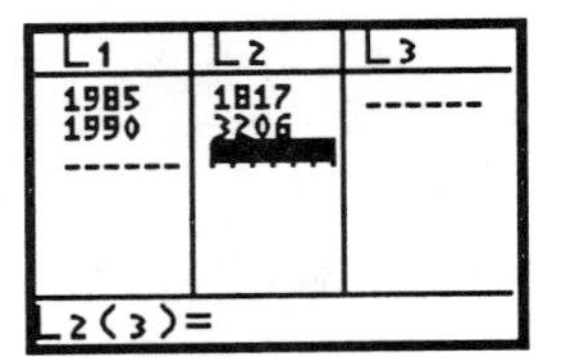

Figure 2. 5

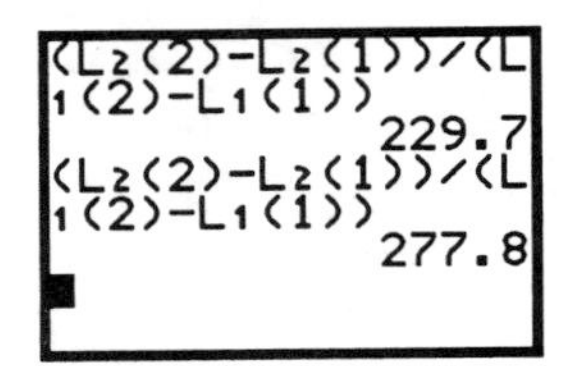

Figure 2. 6

Carefully enter all parentheses.

2.1.2 Working With Longer Lists

Example 3

Below is the Federal debt from 1985 to 1990.

Year	Billions of $
1985	1817
1986	2120
1987	2346
1988	2601
1989	2868
1990	3206

> Note: As previous noted in section 1.1, use the clear shortcut technique to clear lists: [Δ] to L1 [CLEAR] [∇] [▷] [Δ] to L2 [CLEAR] [∇] .

a) Calculate the average rate of change for each year. Enter the above data into L1 and L2. Press [STAT] [1:Edit] . See Figure 2.7.

b) Create the rate of change equation by using the sequence command and store the values to list 3, L2.

The rate of change equation for any year would be :

$$\frac{(L2(N+1) - L2(N))}{(L1(N+1) - L1(N))}$$

Press [2nd] [QUIT] . Type the following:

[2nd] [LIST] [1], select [5:seq( ] . See Figure 2.8.

Type the commands as in Figure 2.9 and store to list 3, L3.

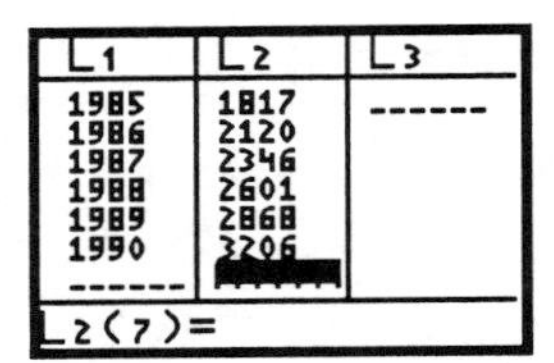

Figure 2. 7

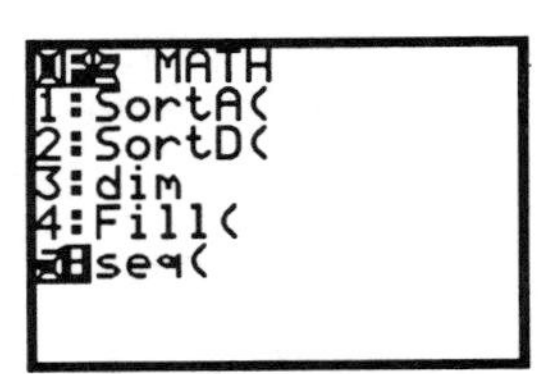

Figure 2. 8

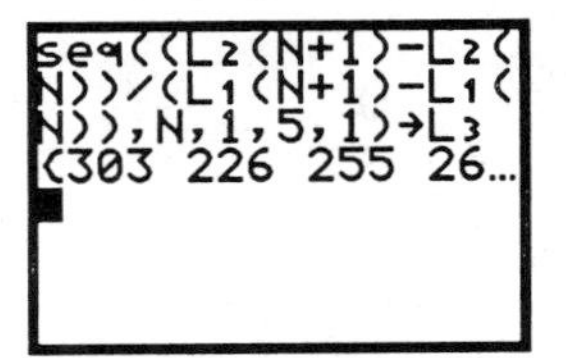

Figure 2. 9

Carefully enter all parentheses

[1] For the TI-83 [2nd] [LIST] <OPS> select [5:seq( ] .

Figure 2.10 shows the rate of change values stored in L3. This is a very handy technique for calculating the rate of change for very long lists.

L1	L2	L3
1985	1817	303
1986	2120	226
1987	2346	255
1988	2601	267
1989	2868	338
1990	3206	------
------	------	

L1(1)=1985

Figure 2. 10

Trouble Shooting: The sequence command generates a list of values by evaluating the *expression*, in terms of a *variable*, from a *begin* value to an *end* value by an *increment* value. Thus the command must include the following:

seq(*expression, variable, begin, end, increment*)

Trouble Shooting: If you want to graph the Rate of Change graph using L1 and L3. The lists must be the same size. Insert a 0 at the beginning of list 3 in the L3(1) position.

Press [2nd] [INS] 0. See Figure 2.11.

L1	L2	L3
1985	1817	0
1986	2120	303
1987	2346	226
1988	2601	255
1989	2868	267
1990	3206	338
------	------	------

L3(1)=0

Figure 2. 11

Insert zero to make lists the same size.

2.2 Special TI-83 Techniques for Finding the Average Rate of Change of Lists

The TI-83 has a shortcut method for finding the rate of change of a list. Calculate the average rate of change (the difference of elements in L2 divided by the difference of elements in L1), then store it to L3 . See Figures 2.12 and 2.13.

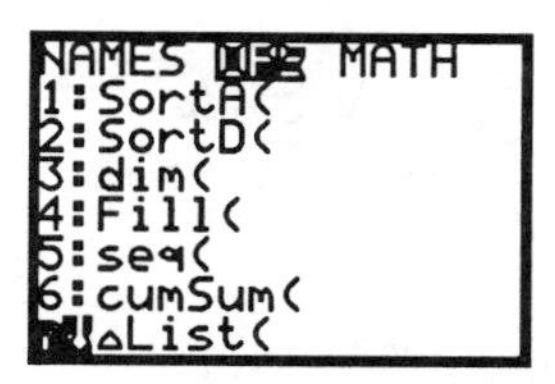

Figure 2. 12 TI-83 Δlist

Press [2nd] [LIST] [▷] to <OPS> , select [7:ΔList( ]. Type as in Figure 2.13.

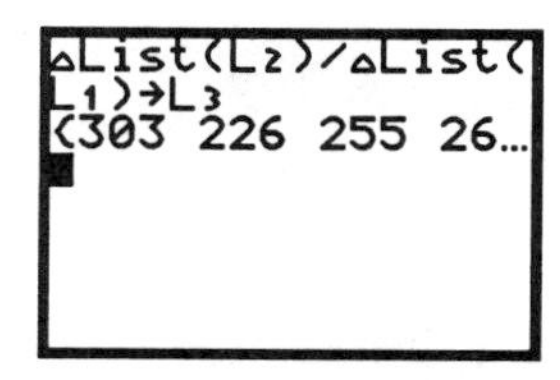

Figure 2. 13 TI-83 Rate of Change

Note: ΔList(*list*) generates a list containing the differences between consecutive elements in a named *list* .

Press [STAT] [1:Edit] , to see all three lists . See Figure 2.14.

L1	L2	L3
1985	1817	303
1986	2120	226
1987	2346	255
1988	2601	267
1989	2868	338
1990	3206	------
------	------	

L3 ={303,226,255...

Figure 2. 14 TI -83 Lists

See the Trouble Shooting box above Figure 2.11 for graphing the Rate of Change.

2.3 Graphing Linear Functions

2.3.1 Definition of a Linear Function

The relationship between the variables x and y is said to be linear i. for any set of points (x, y) the rate of change of y with respect to x is constant. The rate of change is called ***the slope of the line.*** Any collection of points (x, y) that have a linear relationship will satisfy an equation of the form $y = mx + b$ where m is the rate of change of y with respect to x (m = **slope**). b is the y - **intercept** (the value of y when $x = 0$). $y = mx + b$ is called the ***slope intercept form.***

An example of a linear equation (function) is :

$$y = 5x + 4$$

The slope is: $m = 5$
The y-intercept is: $b = 4$

> <u>Note</u>: If needed, refer back to Section 1.10 to review the introduction to graphing.

Example 4

Plot the following points on an $x\,y$ coordinate plane and determine if they lie on a straight line.
(0, 2), (1, 5), (2, 8), (3, 11), (5, 17), and (7, 23)

Organize the data as in Table 2.1

x	0	1	2	3	5	7
y	2	5	8	11	17	23

Table 2. 1

2.3.2 Plotting Points

You can use the graphing calculator to display the data in list form:

1. First CLEAR any data in list 1 (L1) and list 2 (L2).
   Press [STAT] ; select [4:ClrList].
   Press [2nd] [L1] [,] [2nd] [L2] . See Figure 2.15.

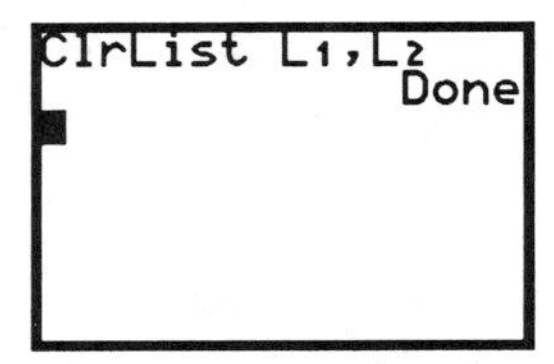

Figure 2. 15

2. Enter the data points. Press [STAT] , select [1:Edit] enter the x- coordinate in L1 pressing [ENTER] after each entry.
   Move the right arrow [▷] to L2 and enter the y-coordinate data . See Figure 2.16.

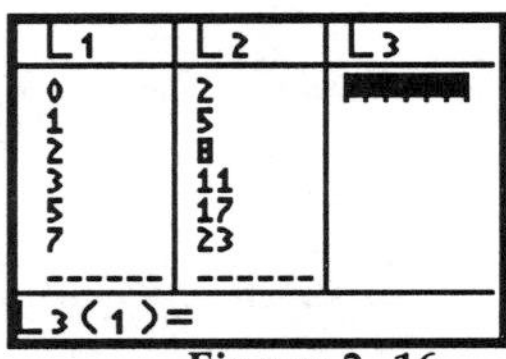

Figure 2. 16

3. Set up the plot. Press [2nd] [STATPLOT] , select [1:Plot 1] as in Figure 2.17 or 2.18.
4. Clear [Y=] and turn off all other plots.

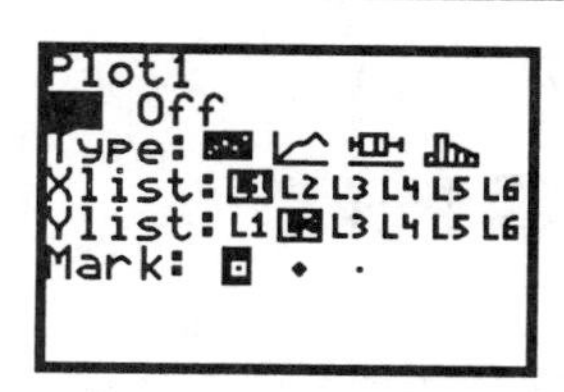

Figure 2. 17 TI-82 Plot 1 screen

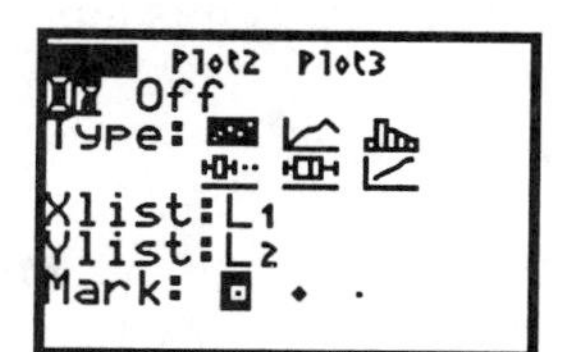

Figure 2. 18 TI-83 Plot 1 screen

To see the plot, set the window based upon the highest and the lowest values of x and y . X: [-1,8] and Y: [-2, 25]. Or let the calculator set it:

Press ZOOM ; select [9:ZoomStat] . See Figure 2.19.

The points appear to lie on a straight line . See Figure 2.20.

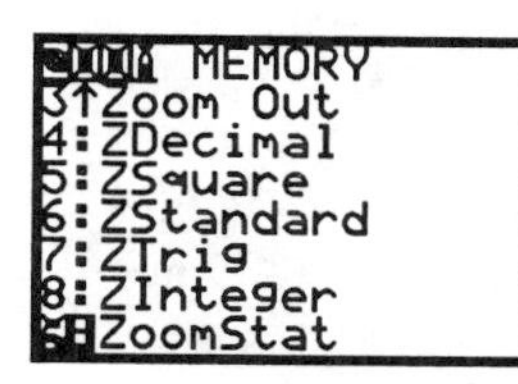

Figure 2. 19

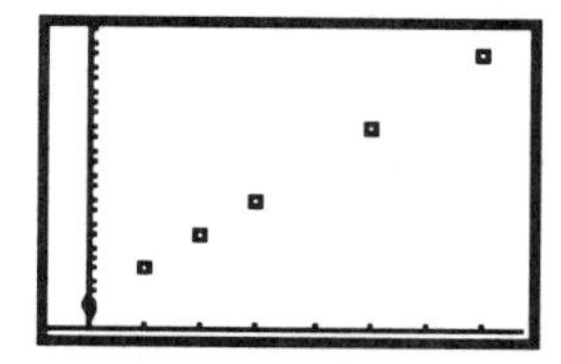

Figure 2. 20

2.3.3 The Algebraic Representation.

You see from the data that when $x = 0$ $y = 2$, so $b = 2$. As x changes one unit, y changes 3 units. The rate of change is 3 units, or $m = 3$. The linear function in slope intercept form $y = mx + b$ is :

$$y = 3x + 2$$

Enter this function into Y1.

Press Y= CLEAR to clear any old functions.

Press 3 X,T,Θ + 2 . See Figure 2.21 or 2.22.

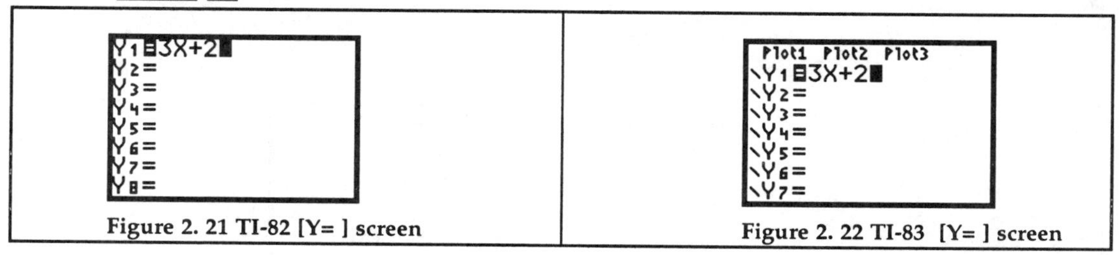

Figure 2. 21 TI-82 [Y=] screen

Figure 2. 22 TI-83 [Y=] screen

Press GRAPH TRACE . See Figure 2.23 or 2.24.

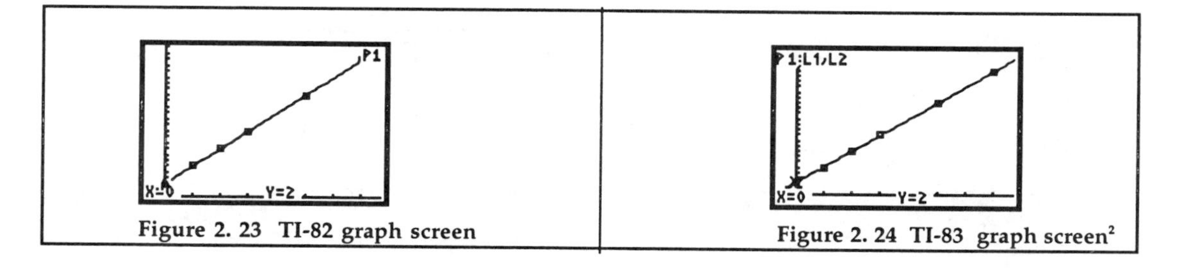

Figure 2. 23 TI-82 graph screen

Figure 2. 24 TI-83 graph screen[2]

[2] The TI-83 graph screen shows the name of the lists being plotted and the equation. From now on only the TI-82 graph screen will be shown.

Verify that the data points are the same as the points on the graph.

Press [2nd] [TBLSET] . Set as in Figure 2.25.

Figure 2. 25

Compare the function values of x and y to the data points in Table 2.1.

Press [2nd] [TABLE] .

X	Y1
0	2
1	5
2	8
3	11
4	14
5	17
6	20

X=0

Figure 2. 26

Figure 2.26 and Table 2.1 share the same values. The difference is that the algebraic model can be used to find other values of y when x is given. For example use [∇] to find $x = 15$, the corresponding value is $y = 47$. See Figure 2.27.

Algebraically;

$$y = 3(15) + 2 = 47$$

X	Y1
13	41
14	44
15	47
16	50
17	53
18	56
19	59

X=15

Figure 2. 27

Use [Δ] to find $x = -10$, the corresponding value is $y = -28$. See Figure 2.28.

Algebraically;

$$y = 3(-10) + 2 = -28$$

X	Y1
-12	-34
-11	-31
-10	-28
-9	-25
-8	-22
-7	-19
-6	-16

X=-10

Figure 2. 28

2.4 Graphing Linear Functions

2.4.1 The ZOOM Menu

The graphing calculator has a default viewing window. This window extends in an x direction from -10 to 10 and in a y direction from -10 to 10.

This is the "standard" window. To set this window quickly;

Press [ZOOM] [6] . See Figure 2.29.

Figure 2. 29

The function and the plot are displayed.

Press [WINDOW] to verify the setting. See Figure 2.30.

Figure 2. 30

Trouble Shooting: Before you continue graphing turn OFF all plots. Press [2nd] [STATPLOT] [4] [ENTER] . After you press ENTER the calculator will tell you *Done* ..

Example 5

Graph the following linear functions in Y1 and Y2:

$f(x) = x + 5$

$g(x) = -2x - 3$

1. Press [Y=] . [CLEAR] all the old entries. Type the functions . See Figure 2.31.
2. Press [GRAPH] . Your graphs should be the same as Figure 2.32.

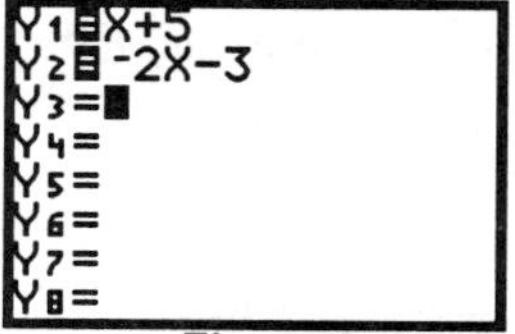

Figure 2. 31

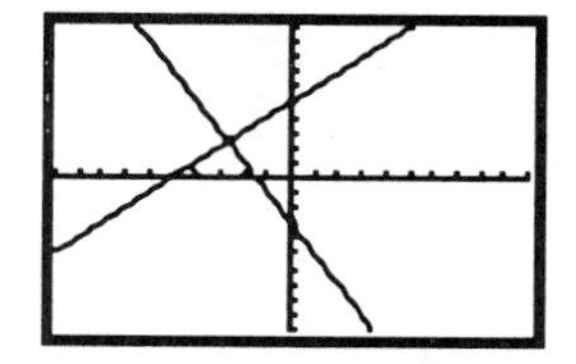

Figure 2. 32

Press [TRACE] then [▷] [◁] to move along the function. Notice x is being incremented in rather ugly decimal values. This is NOT a "friendly" window.

Use [▽] and [Δ] to switch between graphs.

Change the window; press [ZOOM] [8] ; the "integer" window. See Figure 2.33.

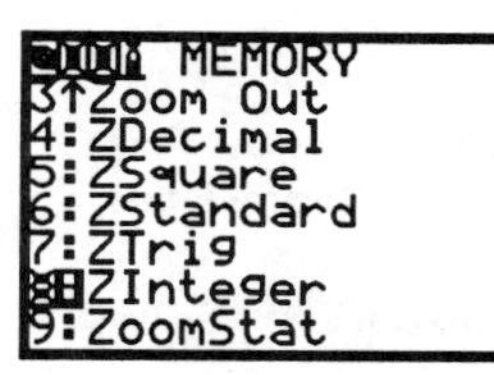

Figure 2. 33

> **Trouble Shooting:** You are now given an opportunity to reposition your cursor, if you should like, before the window changes.
> Press [ENTER] to activate the change.

If you did not move your cursor, your graph will look like Figure 2.34. These are the graphs of the same functions but the window is now much larger. Press [TRACE] and notice that x changes by one integer now.

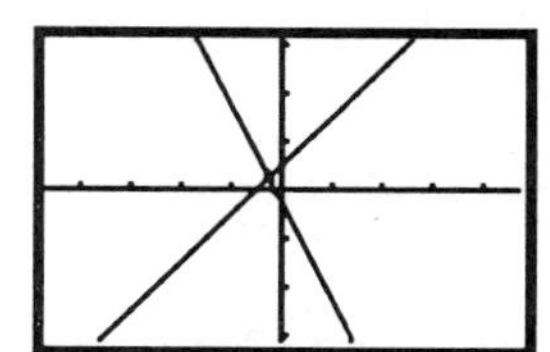

Figure 2. 34

Press [WINDOW] . See Figure 2.35.

Calculate $X_{max} - X_{min}$. The distance is exactly 94, the same as the number of pixel jumps across the screen. This is a "friendly" window. Refer back to Chapter 1 Section 1.10.4 for more on *friendly windows.*

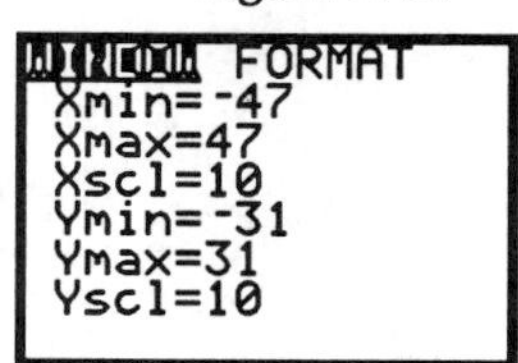

Figure 2. 35
Integer window

Press [ZOOM] [4] . This is the "decimal" window. When you press [TRACE] and move the arrows, x now increments by .1, or one decimal point. This is another "friendly" window . See Figure 2.36.

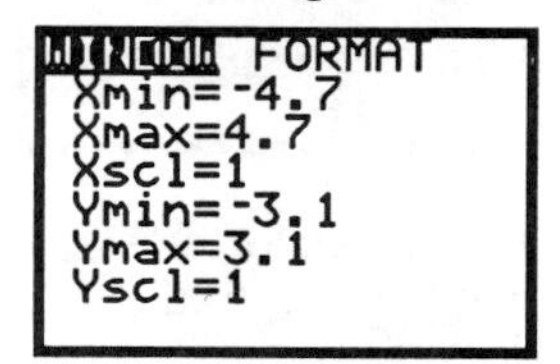

Figure 2. 36
Decimal window

2.4.2 Two Points Determine a Line

If two points (x_1,y_1) and (x_2,y_2) are known, you have enough information to determine the slope or average rate of change using the slope formula:

$$m = \text{slope} = \frac{y_2 - y}{x_2 - x_1}$$

Once the slope is known, you can use $y = mx + b$ to find the equation of the line.

Example 6

If two points on the graph of a linear equation are (5, -1) and (-15, -13), determine the linear equation associated with these points.

1. Find the slope.

$$m = \frac{y_2 - y_1}{x_2 - x_1} = \frac{(-13) - (-1)}{(-15) - (5)} = \frac{-12}{-20} = \frac{3}{5}$$

2. Substitute $m = 3/5$ into $y = mx + b$:
   $y = (3/5)x + b$
3. Pick one of the points for values of x and y. The point (5, -1) means $x = 5$, and $y = -1$. Substitute these values into $y = mx + b$ and solve for b:
   $-1 = (3/5)(5) + b$
   $-1 = 3 + b$
   $b = -4$
4. Write the linear function:

$$y = \frac{3}{5}x - 4$$

This is the linear function through the points (5, -1) and (15, 13) .

2.4.3 Use the Calculator to Verify the Equation of a Line Between Two Points with a Linear Regression Equation

CLEAR L1 and L2. Use a shortcut method:

Press [STAT] [1:Edit] Use [Δ] to position the cursor on the word L1; press [CLEAR] [∇] [▷].

Repeat for L2. See Figure 2.37

Example 7

Find the linear equation that goes through the points (5, -1) and (-15, -13).

1. Enter the points into L1 and L2. See Figure 2.37.
2. Find the linear regression equation.

Press [STAT] [▷] to <CALC> ; select [:LinReg(ax+b)]. See Figure 2.38. Press [2nd] [L1] [,] [2nd] [L2] [ENTER] . See Figure 2.39.

> **Note**: The calculator uses "*a*" instead of "*m*". Both expressions represent a linear function: $y = mx + b = ax + b$

In Figure 2.40 you see the values for the linear equation: a = slope = m = .6 and, b = -4 (the y - intercept).

Thus $y = 0.6x - 4$ is the same as the linear function you determined above:

$$y = (3/5)x - 4 = 0.6x - 4$$

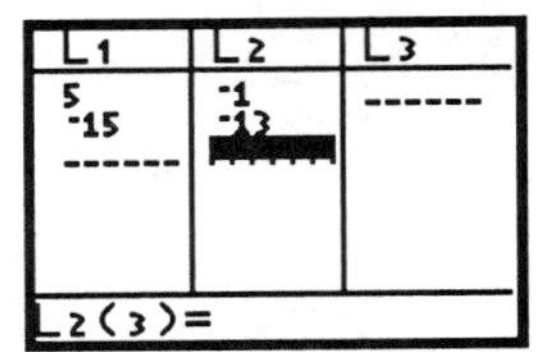

Figure 2. 37

EDIT CALC
1:1-Var Stats
2:2-Var Stats
3:SetUp...
4:Med-Med
5:LinReg(ax+b)
6:QuadReg
7↓CubicReg

Figure 2. 38
On the TI-83 select [4:LinReg(ax+b)]

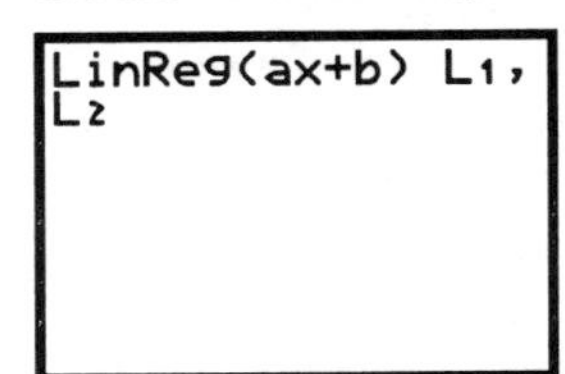

Figure 2. 39

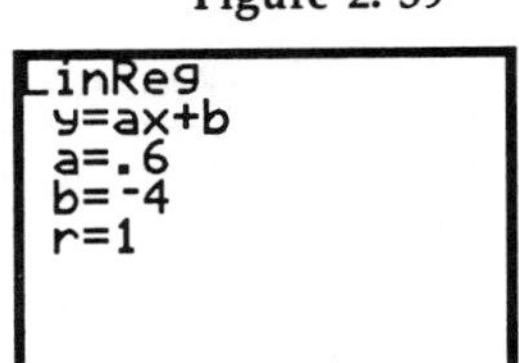

Figure 2. 40
The r value tells you how good of a fit you have. When |r| = 1 the fit is perfect, i.e. r = 1 or r = -1.

2.5 Special lines

> **Trouble Shooting:** Before you continue CLEAR [Y=] and turn OFF all plots. Press [2nd] [STATPLOT] [4] [ENTER]. After you press ENTER the calculator will tell you *Done* ..

2.5.1 Horizontal Lines

Horizontal lines are parallel to the x-axis and have a slope of 0. For this case, if $m = 0$ in $y = mx + b$, the equation becomes

$$y = 0x + \text{b or } y = b.$$

Example 8

Graph the following: Press [Y=]

Y1 = -3
Y2 = 3
Y3 = 2
Y4 = -1

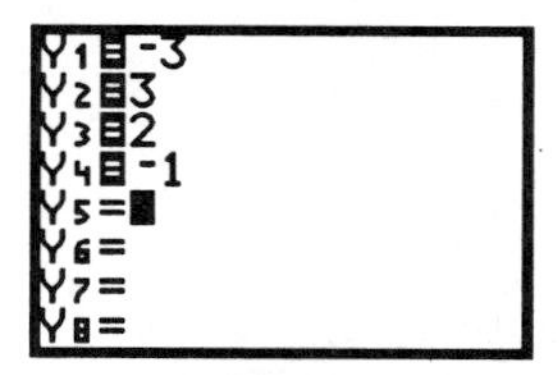

Figure 2. 41

See Figure 2.41. Press [ZOOM] ; select [4:Zdecimal] . See Figure 2.42.

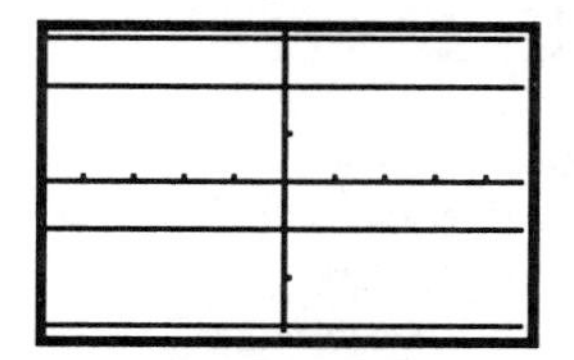

Figure 2. 42

2.5.2 Vertical Lines

Vertical lines are of the form $x = \text{c}$ or x is constant. y can be any number.

Example 9

Draw the vertical lines:

$x = 3, x = -4$ and $x = 1.$

Since vertical lines are *not* functions the calculator must draw them, rather than graph them. The calculator can draw a vertical line using the DRAW menu.

CLEAR [Y=]. Press [2nd] [DRAW] select [4:Vertical]. See Figure 2.43. Use [▷] to position the cursor on $x = 3$. Press [ENTER].

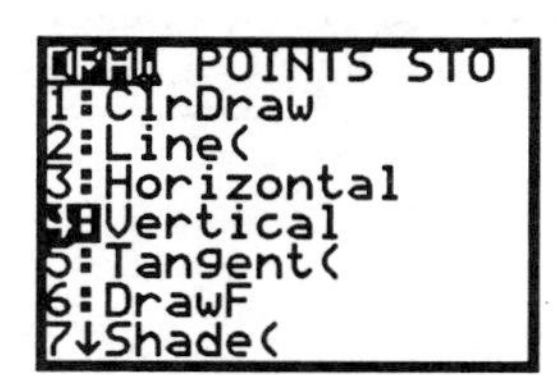

Figure 2. 43

Use [◁] to reposition on $x = 1$. Press [ENTER].

Use [◁] to reposition on $x = -2$. Press [ENTER].

You should see three vertical lines as in Figure 2.44.

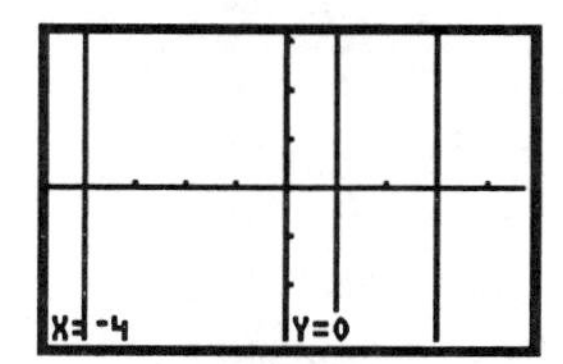

Figure 2. 44

2.3.3 Proportional Relationships

The graph of a line which passes through the origin (0, 0) has a y-intercept = 0. For this case, $b = 0$ or the equation is simply :

$$y = mx \text{ or } y/x = m$$

The ratio of y to x is constant (m) for every point on the graph line.

Example 10

Graph the following direct proportions:

Y1 = 3*x*
Y2= (2/3)*x*
Y3= -4 *x*

All of the graphs in Figure 2.45 go through the origin. Their intercepts are zero. These are examples of proportional linear graphs.

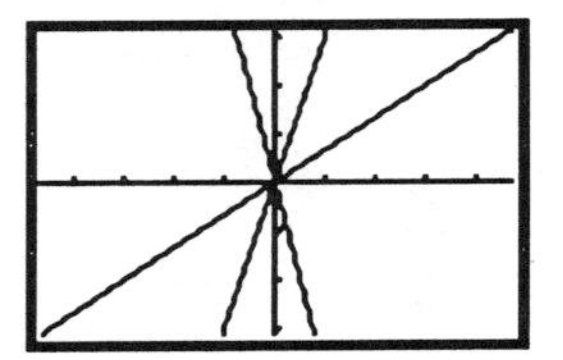

Figure 2. 45

2.5.4 Parallel lines

Parallel lines have the same slope.

Example 11

Graph the following:

Y1= 3*x*
Y2= 3*x* + 4
Y3 = 3*x* - 5
Y4 = 3*x* + 6

See Figure 2.46. Press [ZOOM] ; select [6:Zstandard]. See Figure 2.47. The lines are parallel with different y - intercepts. Press [TRACE] then [▽] to see the intercepts.

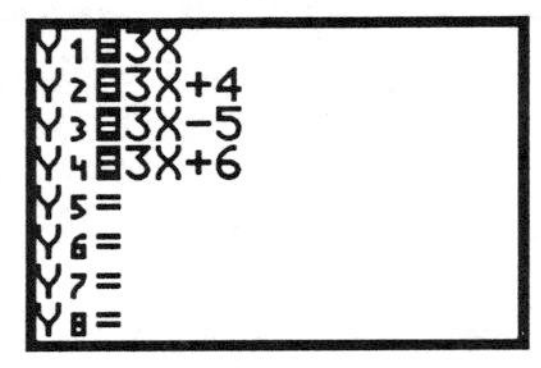

Figure 2. 46

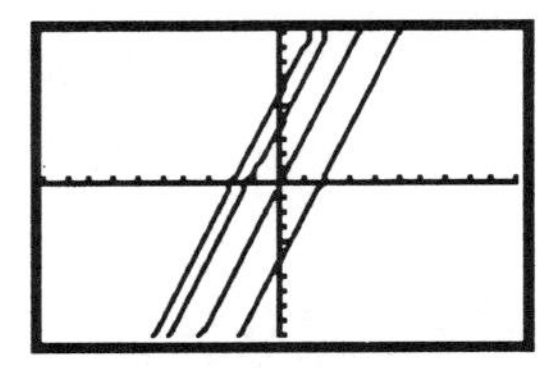

Figure 2. 47

2.3.5 Perpendicular Lines

If the graph of the equation, $y = (2/3)x + 5$ is rotated 90 degrees, the resulting line will have a slope of -3/2. In general, if the slope of a line is m_1, the slope of a perpendicular line is $m_2 = -1/m_1$ (the negative reciprocal of m_1).

> **Note**: For perpendicular lines, l_1 and l_2, the product of their slopes m_1 and m_2 will always equal -1. If $m_1 m_2 = -1$ then l_1 is perpendicular to l_2.

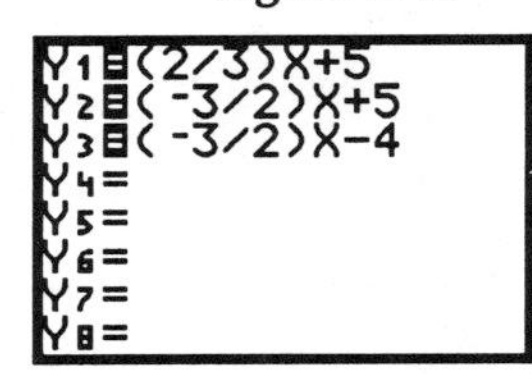

Figure 2. 48

Example 12

Graph:

Y1 = (2/3)*x* +5
Y2= (-3/2)*x* + 5
Y3= (-3/2)*x* - 4

Enter the above functions into [Y=] . See Figure 2.48. Press [GRAPH] . See Figure 2.49. Due to the scaling, the graphs do not look perpendicular. To adjust the window so that the x and y scaling are proportional; press [ZOOM] ; select [5:Zsquare]. See Figure 2.50. Now the graphs look perpendicular. See Figure 2.51.

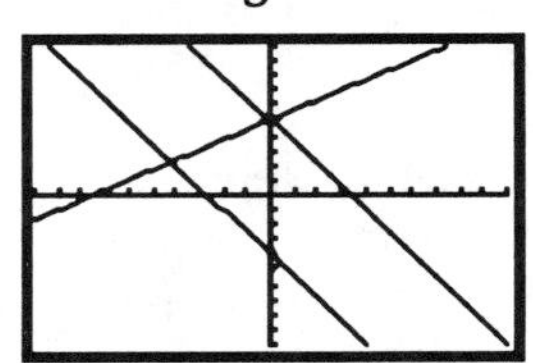

Figure 2. 49

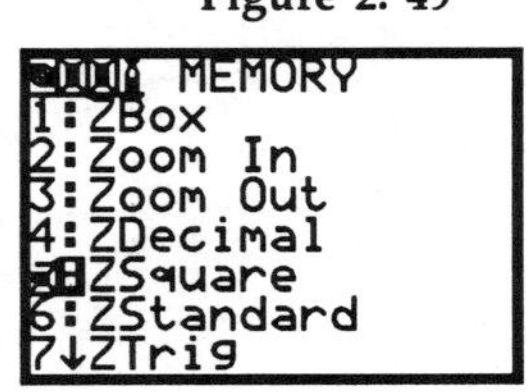

Figure 2. 50

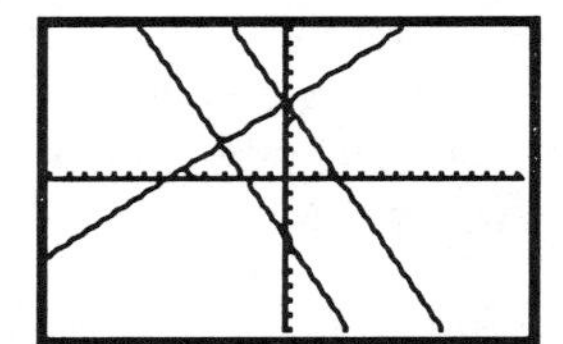

Figure 2. 51

> **Note**: Both Y2 and Y3 are perpendicular to Y1. Y2 is parallel to Y3 because their slopes are equal.

2.5.6 Entering Linear Equations Not in Function Form

The equation $y = mx + b$ is called the slope-intercept form of a linear equation, or the function form, since the equation is solved for y. A linear relationship may also be described by an equation in the *standard form*:

$$Ax + By = C.$$

Example 13

Transform the equation $-2x + 3y = 15$ into function form so that you can enter it into your calculator and graph it.

$-2x + 3y = 15$	
$3y = 2x + 15$	add $2x$
$y = (2/3)x + 15/3$	divide by 3
$y = (2/3)x + 5$	simplify

$-2x + 3y = 15$ is equivalent to $y = (2/3)x + 5$ which you graphed above in example 12. Refer back to Figures 2.48 to see how the function was entered. See Figure 2.52 for the graph of $y = (2/3)x + 5$ by itself.

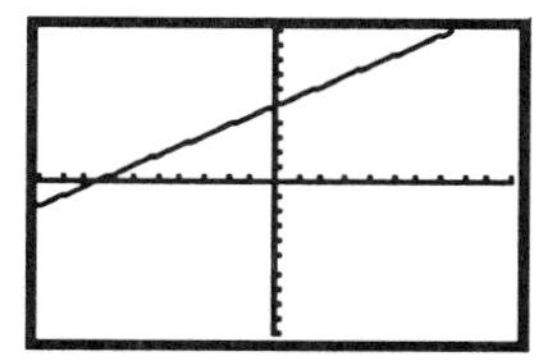

Figure 2. 52
The graph of Y1 = $(2/3)x + 5$

2.3.5 Fractional Coefficients of Linear Functions

Notice that in all of the examples above when the coefficient of x was a fraction, you enclosed the fraction in parentheses. See what happens if you do not use parentheses. The TI-82 and TI-83 behave differently. Enter the function $y = 1/2x$.

Press ZOOM 4 .

The TI-82 interprets this as $y=1/(2x)$. See Figure 2.53. While the TI-83 interprets the function as $y=(1/2)(x)$ as in Figure 2.54. **<u>As a precaution always put fractions in parentheses</u>**.

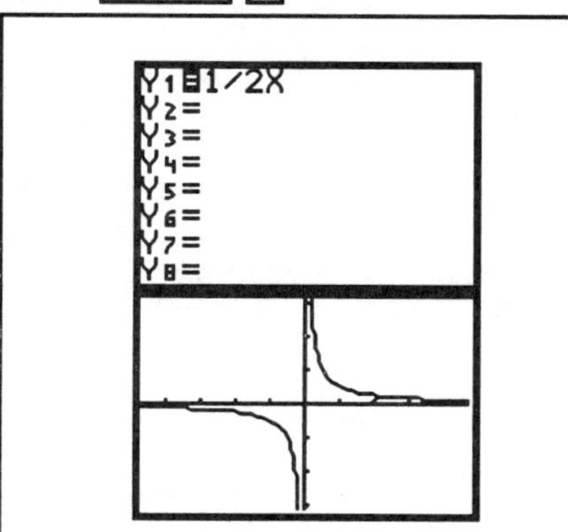

Figure 2. 53 TI-82 NOT a Linear graph.

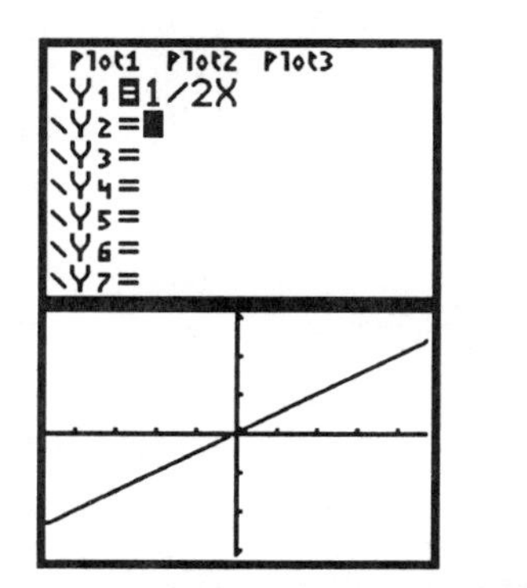

Figure 2. 54 TI-83 linear graph.

Extended Exploration 1: Education vs. Income Using Linear Regression Equations

EE1.1 Linking Calculators

This course comes with programs that store data to the graphing calculator. Once the data is stored to lists, you can create plots and find regression equations on the data. Your instructor will be able to download the programs from a computer to a calculator. Your calculator came with a cable that allows you to link calculators. Once linked, you can receive and send programs as well as data lists.

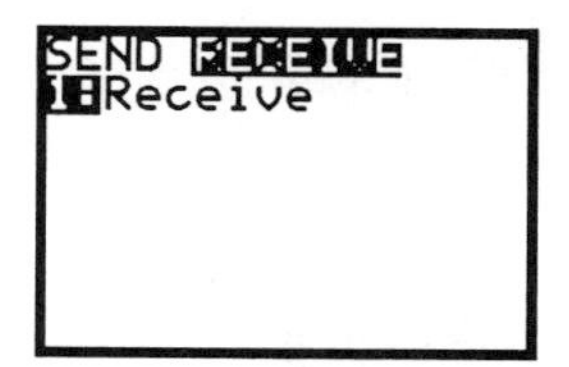

Figure EE1. 1

Figure EE1. 2

EE1.1.1 Receiving Data

1. Attach the cable to both calculators. Be sure to push the cable **all** the way in.
2. Press [2nd] [LINK] [▷] to <RECEIVE>.

   Press [ENTER] . See Figure EE1.1.
3. The receiving calculator must say ***Waiting...***

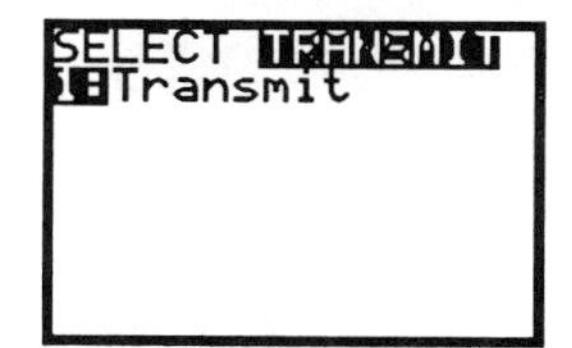

Figure EE1. 3

EE1.1.2 Sending data:

1. Press [2nd] [LINK] ; select [3:Select Current...].
2. [▽] to the programs or lists to be sent.

   Press [ENTER] to select. A small square indicates the selection has been made . See Figure EE1.2.
3. [▷] to <TRANSMIT> press [ENTER] . See Figure EE1.3.
4. Wait for the message ***Done..*** on the receiving calculator . See Figure EE1.4.

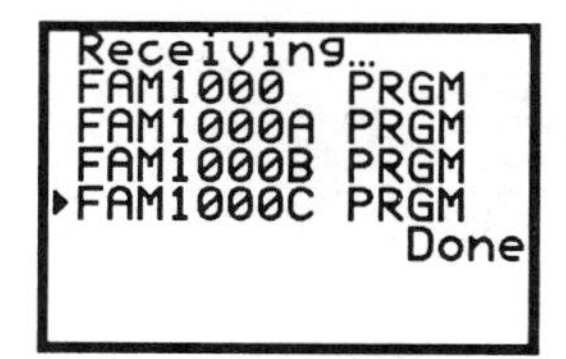

Figure EE1. 4

Figure EE1. 5

EE1.1.3 Running a Program

Press [PRGM] , then select the program.

Press [ENTER] . See Figures EE1.5 and EE1.6.

The program has stored the FAM1000 data to List1 though List6.

Press [STAT] select [1:Edit] to see the data in the lists . See Figure EE1.7.

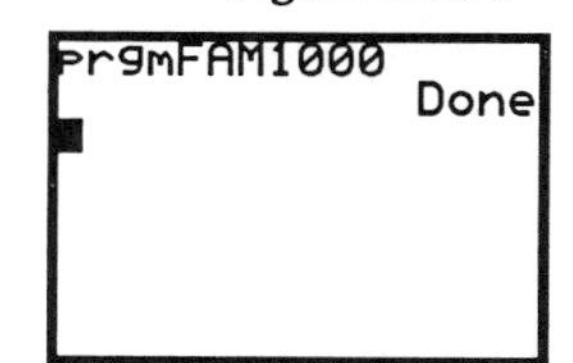

Figure EE1. 6

To see the program press [PRGM] [▷] to <EDIT>. Select the program to view. Press [ENTER]. See Figure EE1.8.

L1	L2	L3
0	12200	0
1	15000	1
2	11134	2
3	8333	5
5	8583	6
6	15041	7
7	17112	8

L1(1)=0

Figure EE1. 7

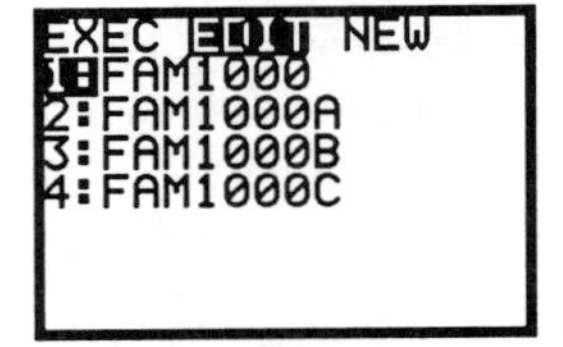

Figure EE1. 8

The program will look like Figure EE1.9. At the end of this chapter the complete FAM 1000 data sets will be listed with appropriate column headings.

Trouble shooting: If you get lost and don't know which menu you are in, press [2nd] [QUIT] and start over.

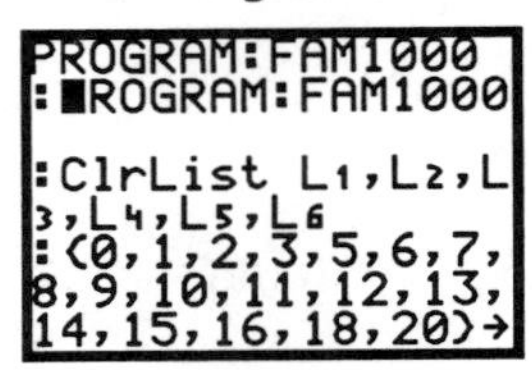

Figure EE1. 9

EE1.2 Linear Regression Equations

The FAM 1000A program stores the years of education in L1 and the mean personal income in L2. To plot the data press [2nd] [STATPLOT] [ENTER]. See Figure EE1.10.

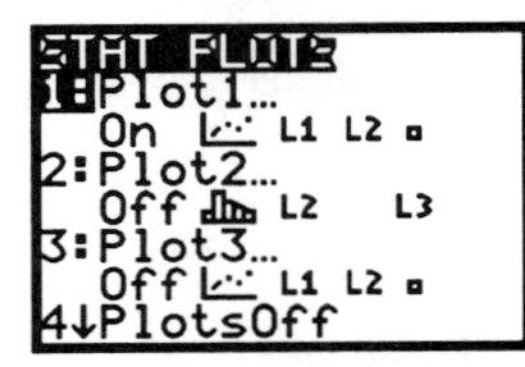

Figure EE1. 10

EE1.2.1 Set up the plot

1. Select **ON** [ENTER].
2. Select Type: **scatterplot icon,** [ENTER].
3. Select Xlist: **L1** [ENTER].
4. Select Ylist: **L2** [ENTER].
5. Select Mark: ▫ . See Figure EE1.11.

Figure EE1. 11

Trouble shooting:
CLEAR [Y=] and turn **off** all other plots.

EE1.2.2 Draw a Scatter Plot

Press [ZOOM]; select [9:ZoomStat]. See Figure EE1.12 and EE1.13. Review Chapter 1 Section 1.10.5 for selecting appropriate viewing windows.

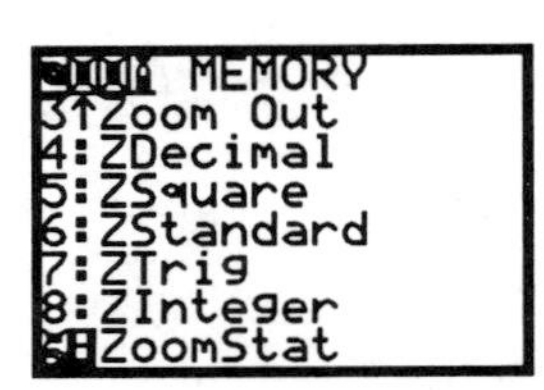

Figure EE1. 12

EE1.2.3 Find the Equation

Press [STAT] [▷] to <CALC>, select [:LinReg(ax+b)] [2nd] [L1] [,] [2nd] [L2] [ENTER]. See Figures EE1.14 through EE1.16.

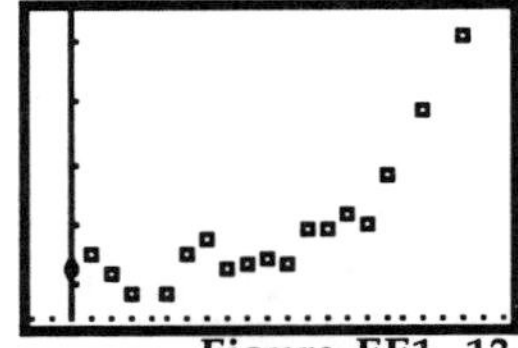

Figure EE1. 13

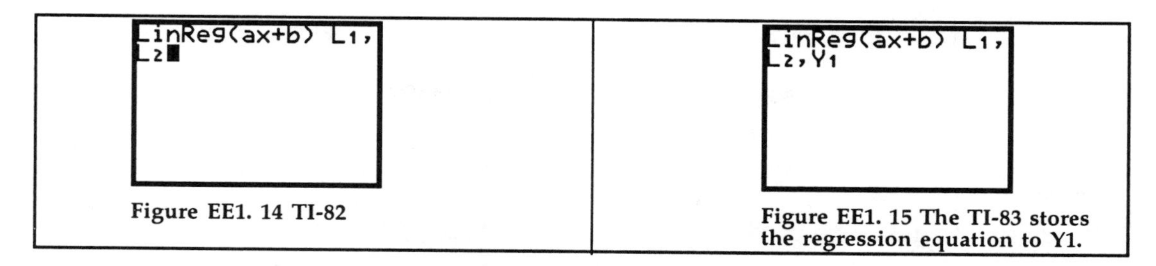

Figure EE1. 14 TI-82

Figure EE1. 15 The TI-83 stores the regression equation to Y1.

Enter the equation into Y1. For the TI-82 press [Y=] [VARS] ; select [5:statistics] [▷] [▷] to <EQ> ; select [:RegEQ] . See Figure EE1.17.

> **Note** : Figure EE1.15 shows how the TI-83 stores the equation into Y1 and Figure EE1.16 shows the TI-83 regression equation screen.

Recall $a = m$, the slope or rate of change and b = the y-intercept. Press [GRAPH] .

Figure EE1.18 shows both the scatterplot and the graph of the regression equation.

The equation means that on average for every year of education a person earns an additional $1450.02 and with no education a person earns only $4999.72 . This corresponds to the slope of the equation, which is 1450.02, and the y-intercept which is 4999.72.

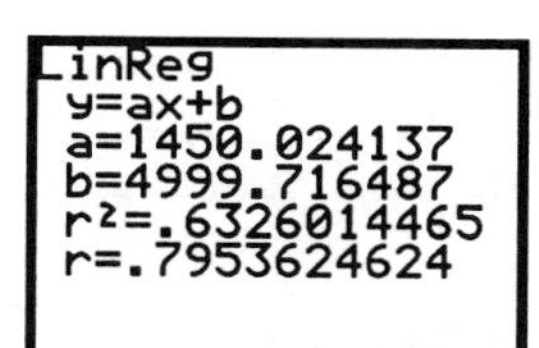

Figure EE1. 16 TI-83 screen

The TI-83 linear regression equation. The r value tells how good of a fit you have. To see the r values: press [2nd] [CATALOGUE] choose DiagnosticsOn.

Figure EE1. 17

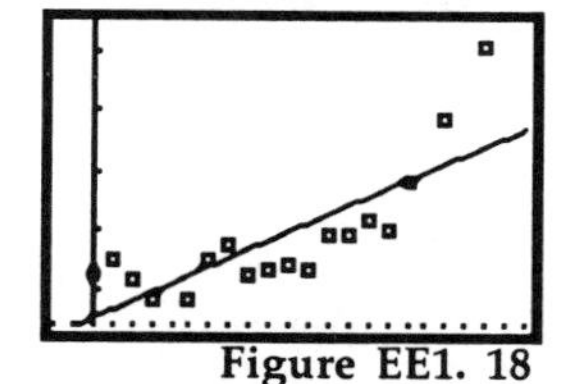

Figure EE1. 18

EE1.3 TI-83 Naming Lists

The TI-83 allows you to name lists and store them. Press [2nd] [L1] [STO▷] [2nd] [ALPHA] then type no more that 5 letters. Repeat for list 2 , L2 . See Figure EE1.19). To place the list in the edit screen press [STAT] ; select [5:SetUpEditor], [2nd] [LIST] then choose the names, [ENTER] . See Figure EE1.19. Press [STAT] [1:Edit]to see the named lists . See Figure EE1.20.

> Revert back to the L1-L6 edit screen; press [STAT] ; select [5:SetUpEditor] [ENTER] .

Figure EE1. 19

EDUC	INCME	----- 1
0	12200	
1	15000	
2	11134	
3	8333	
5	8583	
6	15041	
7	17112	

EDUC(1) =0

Figure EE1. 20

This data is also on the Website www.wiley.com/Kime??????

FAM 1000A: Mean Personal Wages			
Years of Education	Mean Personal Wages ($) (all)	Mean Personal Wages ($) (men)	Mean Personal Wages ($) (women)
L1	L2	L3	L4
1	$7,328	$8,603	$3,500
4	$9,462	$10,500	$8,869
6	$14,932	$17,898	$9,000
8	$13,695	$16,241	$9,558
9	$12,811	$22,910	$7,201
10	$18,319	$22,046	$14,281
11	$12,997	$16,739	$8,319
12	$22,303	$26,599	$17,654
14	$27,756	$36,512	$18,737
16	$40,335	$52,076	$26,217
18	$58,610	$69,262	$40,502
20	$84,528	$99,399	$39,916

FAM 1000B: Mean Personal Total Income			
Years of Education	Mean Personal Total Income ($) (all)	Mean Personal Total Income ($) (men)	Mean Personal Total Income ($) (women)
L1	L2	L3	L4
1	$8,425	$10,067	$3,500
4	$11,289	$11,881	$10,950
6	$14,932	$17,898	$9,000
8	$17,609	$19,555	$14,446
9	$15,243	$23,990	$10,383
10	$20,903	$24,848	$16,630
11	$15,580	$19,036	$11,261
12	$26,411	$32,336	$19,997
14	$31,553	$40,365	$22,475
16	$49,091	$63,198	$32,127
18	$68,391	$79,781	$49,029
20	$104,351	$124,337	$44,395

This data is available as a GraphLink File for the TI-82 or TI-83, and as an Excel file, available at: http://www.wiley.com/college/Kimeclark.

FAM 1000C: Mean Personal Total Income for White and Nonwhite Men			
Years of Education	Mean Personal Total Income ($) All Men	Mean Personal Total Income ($) All White Men	Mean Personal Total Income ($) All Nonwhite Men
L1	L2	L3	L4
1	$10,067	$10,067	nul
4	$11,881	$11,881	nul
6	$17,898	$17,278	$21,000
8	$19,555	$20,677	$15,815
9	$23,990	$21,763	$32,899
10	$24,848	$26,669	$3,000
11	$19,036	$20,162	$15,658
12	$32,336	$33,510	$25,795
14	$40,365	$41,454	$34,672
16	$63,198	$63,568	$60,060
18	$79,781	$79,992	$72,828
20	$124,337	$121,780	$139,676

FAM 1000D: Mean Personal Total Income for White and Nonwhite Women			
Years of Education	Mean Personal Total Income ($) All Women	Mean Personal Total Income ($) All White Women	Mean Personal Total Income ($) All Nonwhite Women
L1	L2	L3	L4
1	$3,500	$3,500	nul
4	$10,950	$11,415	$8,161
6	$9,000	$9,000	nul
8	$14,446	$10,989	$24,818
9	$10,383	$9,634	$13,004
10	$16,630	$15,757	$19,251
11	$11,261	$9,955	$14,134
12	$19,997	$20,515	$17,255
14	$22,475	$23,558	$18,758
16	$32,127	$30,984	$43,875
18	$49,029	$50,309	$37,500
20	$44,395	$34,700	$57,320

Chapter 3
Linear Systems and Piecewise Functions

3.1 Intersecting Lines

Lines that cross each other are said to ***intersect***. All lines eventually intersect unless they are parallel lines or they are the same line. Finding the solution to a system of linear equations is finding the point of intersection.

Example 1

Graph the following system of linear equations and determine the point of intersection, if it exists.

$$f(x) = 3x + 4 \quad \text{and} \quad g(x) = 0.5x - 1$$

Press [Y=] [CLEAR] to clear all the old functions. Enter the above equations into Y1 and Y2 . See Figure 3.1.

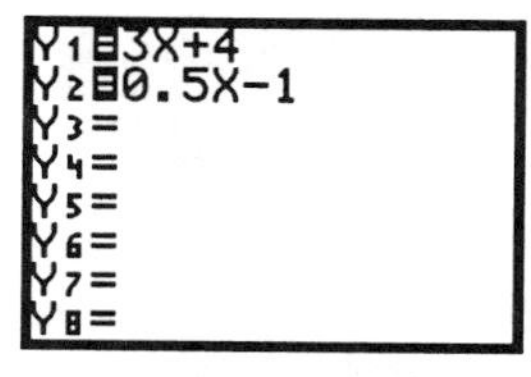

Figure 3. 1

Press [ZOOM] ; select [6: ZStandard] . Press [TRACE] then use [▽] or [△] to estimate the point of intersection. It looks like the graphs cross when x is about -2 and y is about -2 . See Figure 3.2.

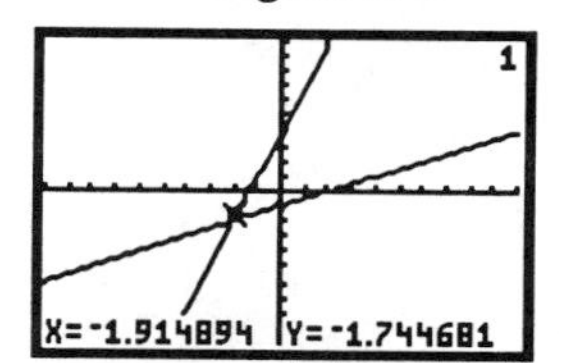

Figure 3. 2

3.1.1 The Calculate Menu

Use the calculator to find the point of intersection.

Press [2nd] [CALC] , select [5:intersect] . See Figure 3.3.

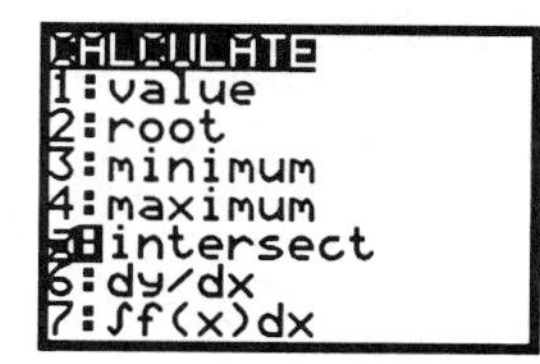

Figure 3. 3

The calculator prompts you:

1. Select the first curve, press [ENTER] .
2. Select the second curve, press [ENTER] .
3. Using the arrows move the cursor near the point of intersection (your guess).

   Press [ENTER] . See Figures 3.4 and 3.5.

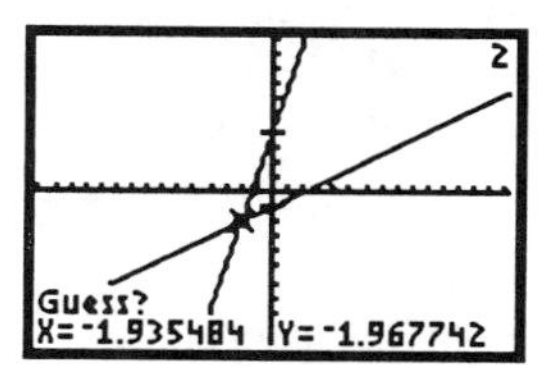

Figure 3. 4

The point of intersection is (-2,-2).

3.1.2 Algebraic Verification

Check algebraically:

If $x = -2$ then

Y1= $3x + 4$ becomes Y1 = 3(-2) + 4 = -2

Y2= $0.5x$ -1 becomes Y2 = 0.5(-2) - 1 = -2

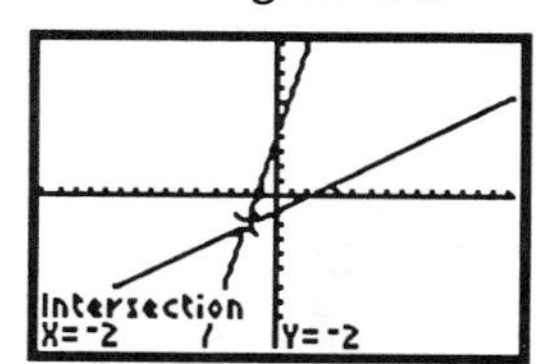

Figure 3. 5

Both equations have the same y values so the solution is $x = -2$ and y=-2 or the point of intersection is (-2,-2).

3.1.3 Solving Algebraically

Since Y1=Y2 at the point of intersection, you solve algebraically by substitution:

$3x + 4$	$= 0.5\,x - 1$	
$2.5\,x + 4$	$= -1$	add $-0.5\,x$
$2.5\,x$	$= -5$	add -4
x	$= -5/2.5 = -2$	divide by 2.5

Substitute x =-2 in Y1 and Y2 as above to find y = -2. The solution is (-2, -2).

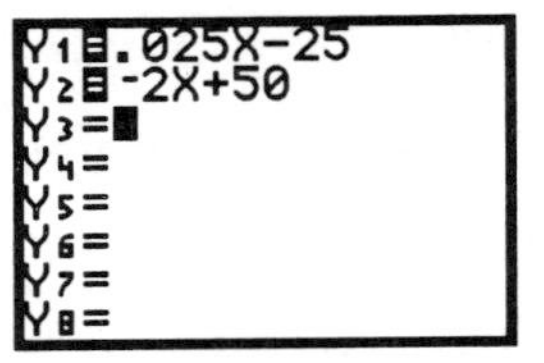

Figure 3. 6

3.1.4 Adjusting the Window

Example 2

Find the point of intersection for the system of equations :

h(x) = 0.025x - 25 and

k(x) = -2x + 50

Enter the expressions into Y1 and Y2. Press GRAPH . See Figures 3.6 and 3.7.

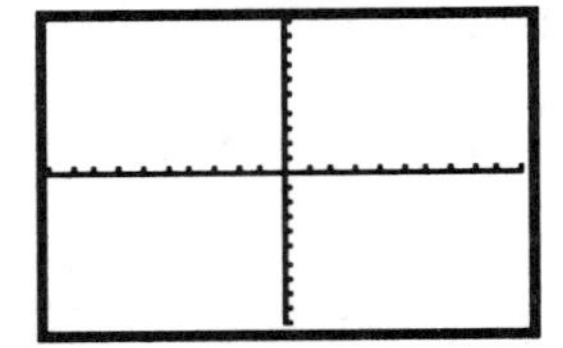

Figure 3. 7

The graphs do not appear! Where are they? Press TRACE . This will give you points on the graph . See Figure 3.8. Using algebra, let x = 0 to find the y - intercepts for the graphs at (0, -25) and (0, 50). To see these points we must choose Ymin and Ymax beyond those points. Set the WINDOW to Figure 3.9. Now we see parts of the graphs. The point of intersection appears to be further to the right . See Figure 3.10.

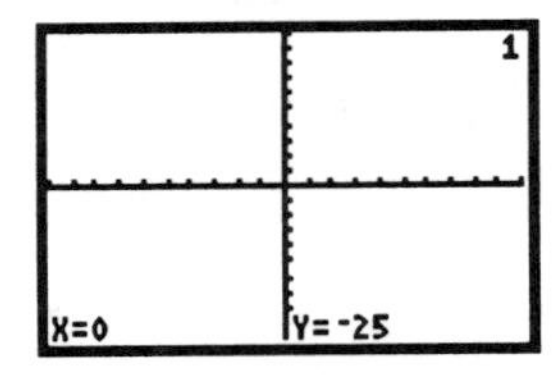

Figure 3. 8

WINDOW FORMAT
Xmin=-10
Xmax=10
Xscl=1
Ymin=-60
Ymax=60
Yscl=10

Figure 3. 9

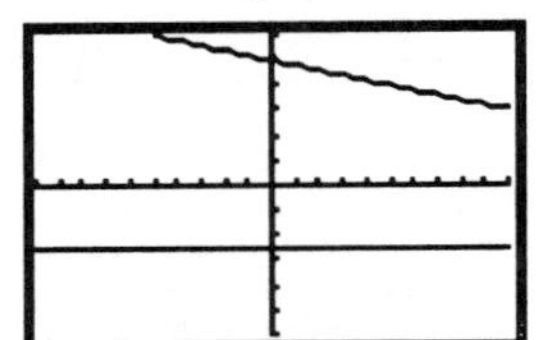

Figure 3. 10

Press 2nd TABLE for more information. Y1 is climbing very slowly (m = 0.025) while Y2 is decreasing and somewhere around x = 37 they are about equal . See Figure 3.11.

X	Y1	Y2
33	-24.18	-16
34	-24.15	-18
35	-24.13	-20
36	-24.1	-22
37	-24.08	-24
38	-24.05	-26
39	-24.03	-28

X=37

Figure 3. 11

Experiment with different window settings. Set Xmax so that the point of intersection and beyond can be seen. Press GRAPH and repeat the CALC Menu steps in Section 3.1.1. See Figure 3.12.

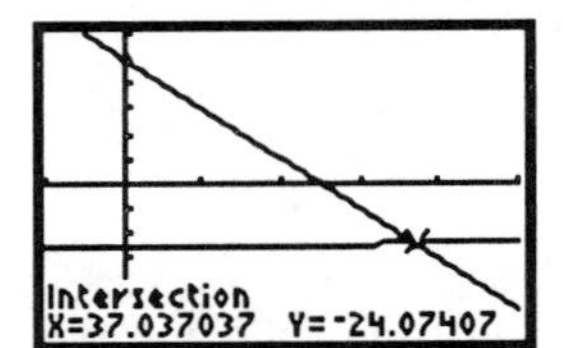

Figure 3. 12

3.1.5 Solve Algebraically to Confirm

Let Y1=Y2:

$$0.025x - 25 = -2x + 50$$
$$2.025x - 25 = 50$$
$$2.025x = 75$$
$$x = 75/2.025 = 37.0370370...$$

Substitute the x value into Y1 and Y2. To find the value of y.

Y1= 0.025(75/2.025) - 25 =-24.07407407...

Y2= -2(75/2.025) + 50 = -24.07407407...

Although the algebra is pretty fast the arithmetic could still use some help from a calculator!

3.2 Piecewise Functions

The graphing calculator can graph piecewise functions very easily. However, it is important to understand how the graphing calculator performs a test.

3.2.1 The TEST Menu

The graphing calculator can tell if a statement is true or false using the TEST menu. Notice that you find the equal and inequality symbols here. Press [2nd] [TEST]. See Figure 3.13. Go to the Home Screen.

Press  [2nd] [QUIT].

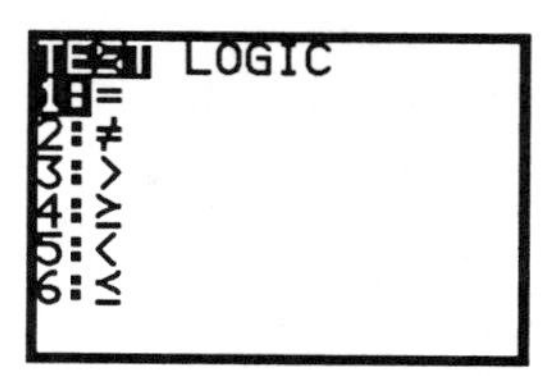

Figure 3. 13

Example 3

Determine if the following are True or False by typing the following:

a. $5 = 5$

b. $5 \leq 7$

c. $5 = 4$

d. $5 \geq 7$

See Figures 3.14 and 3.15.

> **Note**: The calculator is performing a test. It tells you **1** for **True** and **0** for **False**.

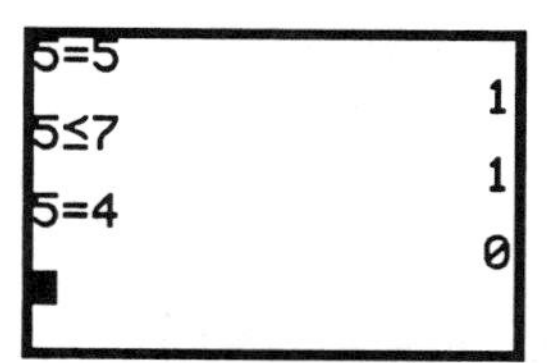

Figure 3. 14

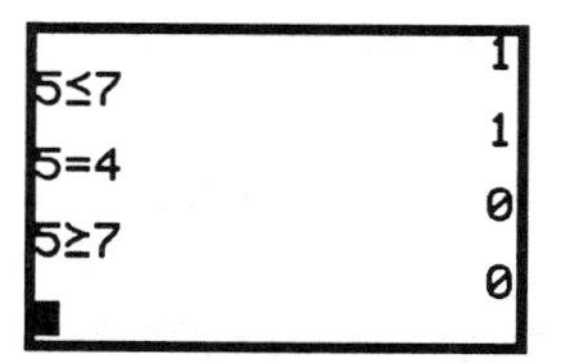

Figure 3. 15

This method can be used to enter a piecewise function (a function defined under several conditions) into the calculator.

Example 4

Graph the piecewise function:

$$f(x) = \quad y = \begin{cases} x + 2 \text{ for } x > 0 \\ -x - 3 \text{ for } x \leq 0 \end{cases}$$

y is defined under two conditions:

1. when $x > 0$, use $y = x + 2$
2. when $x \leq 0$, use $y = -x - 3$

Press [Y=] and enter the equations as in Figure 3.16 .

Since this is a piecewise function, if $x > 0$ choose the function $y1 = x + 2$ but if $x \leq 0$ choose the function $y2 = -x - 3$. See Figure 3.17.

> **Note**: On the calculator we will use the division sign to enter the *x* value condition. The calculator will perform a test by putting 1 or 0 in the denominator. When the denominator is 0 the function is undefined and no points will be drawn. When the denominator is 1 the function is plotted.

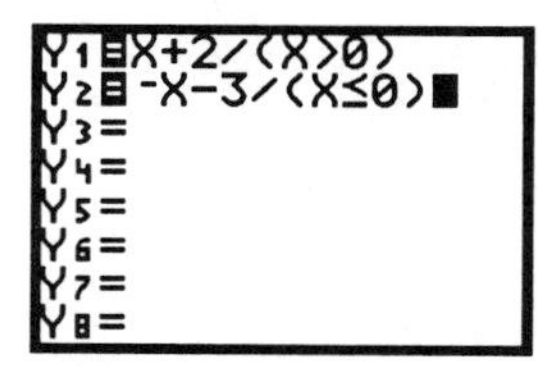

Figure 3. 16

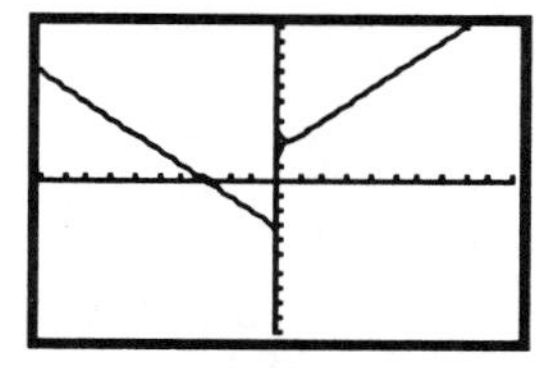

Figure 3. 17

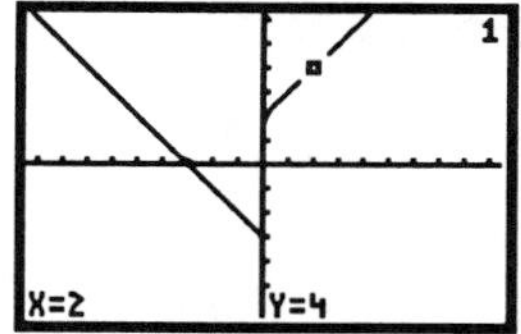

Figure 3. 18

The graph of $f(x)$ is in two pieces.
When you are on Y1 you are on the graph of condition one, or $f(x) = x + 2$.
Press [TRACE]. See Figure 3.18.

When you are on Y2 you are on the graph of condition two, or $f(x) = -x - 3$.
Press [▽], then [TRACE] [◁], to duplicate Figure 3.19. Notice that when the condition no longer applies, no y value is given. For example if $x = 3$ condition two no longer applies. See Figure 3.20.

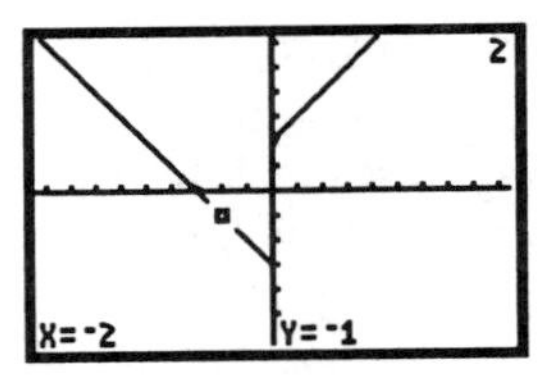

Figure 3. 19

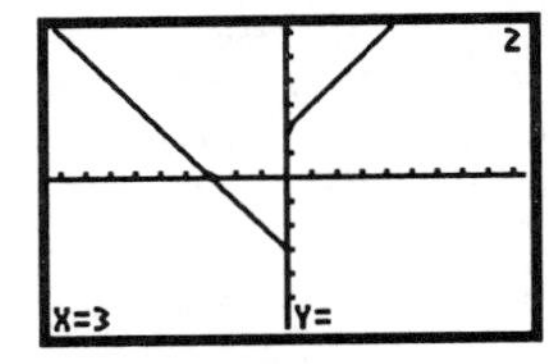

Figure 3. 20

To find the value of y for $x = 3$, you must switch to condition one. [Δ] to the graph of Y1. See Figure 3.21.

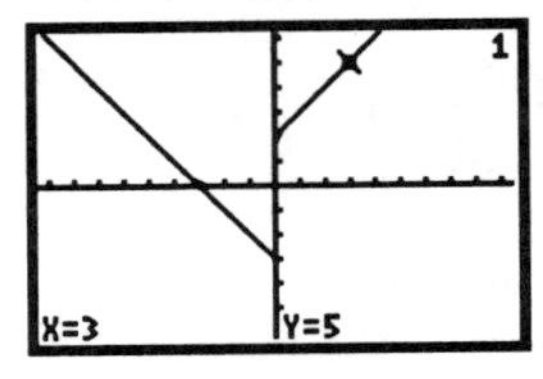

Figure 3. 21

3.2.2 An Alternate Method for Graphing Piecewise Functions

Many people choose to write a piecewise function all on the same line since $f(x)$ is defined for all x. This entails making an adjustment to the mode. Press [MODE]; select **Dot** [ENTER]. See Figure 3.22.

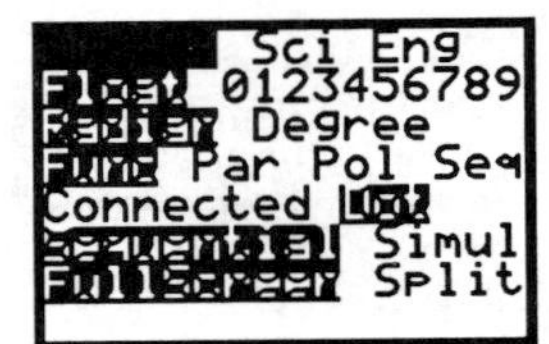

Figure 3. 22

Example 5

Graph the piecewise function

$$f(x) = \begin{cases} (x+3)^2 - 5 & \text{for } x < -2 \\ 2x + 6 & \text{for } x \geq -2 \end{cases}$$

Press [Y=] [CLEAR] to clear all expressions. Type the piecewise function commands as in Figure 3.23. Press [GRAPH]. See Figure 3.24.

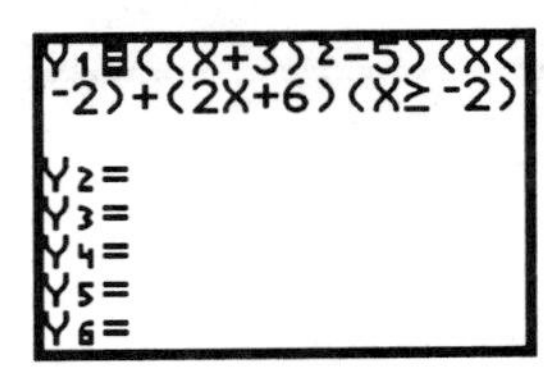

Figure 3. 23 See Troubleshooting.

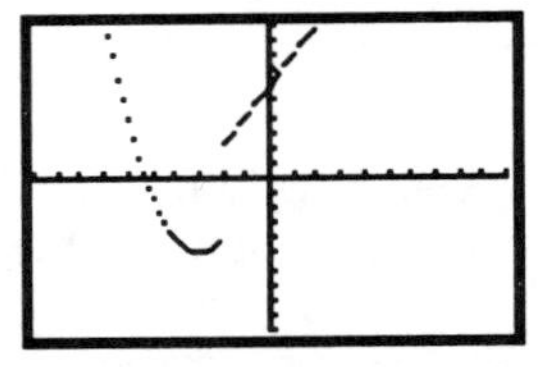

Figure 3. 24 Graph in Dot mode.

The advantage of this method is that you can TRACE on the function in the normal manner.

Troubleshooting:

1. If you are in connected MODE the calculator tries to connect the point from the end of one piece of the graph to the end of the other piece, giving the false impression that the function is continuous. See Figure 3.25.
2. Since you use multiplication for the test, the calculator places a 1 or a 0 inside the test parentheses. This sometimes gives the false value of 0 rather than an undefined value for the function.

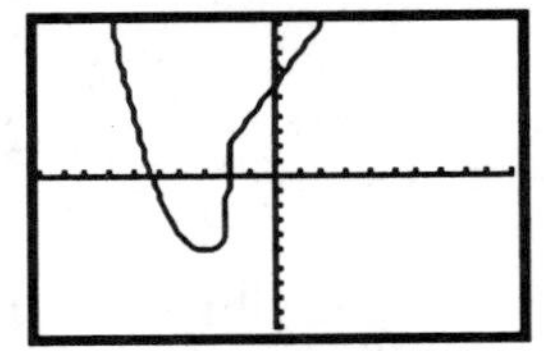

Figure 3. 25
False Graph in Connected mode.

Note: Change back to connected MODE.

Example 6

An 8% flat income tax is represented by *f(x)*. Under the flat tax everyone pays 8%, regardless of how much money is earned per year. A graduated income tax is represented by the piecewise function *g(x)*. Under this plan the first $20,000 is tax free, then between $20,000 and $100,000, 5% tax is paid. If you make beyond $100,000 you pay a 10% tax. Graph these functions.

$$f(x) = .08x \quad \text{for } x \geq 0$$

$$g(x) = \begin{cases} 0 & \text{for } 0 \leq x \leq 20000 \\ .05x(x - 20000) & \text{for } 20000 < x \leq 100000 \\ 4000 + .10(x - 100000) & \text{for } x > 100000 \end{cases}$$

Enter the piecewise functions into the graphing calculator as individual functions with separate conditions. Enter the function as shown in Figure 3.26.

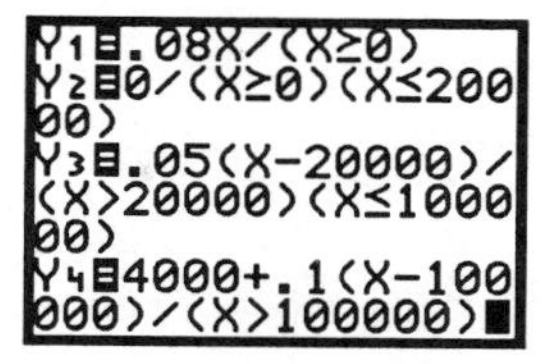

Figure 3. 26

Y1 = $0.08x$ /($x \geq 0$)

Y2 = $0/(x \geq 0)(x \leq 20000)$

Y3 = $0.05(x - 20000) / (x > 20000)(x \leq 100000)$

Y4 = $4000 + .1(x - 100000) / (x > 100000)$

3.2.3 Adjust the Viewing Window

Look at the values for x. They go beyond $100,000. Evaluate Y1 and Y4 for x =110,000 See Figure 3.27. With this new information about x and y set the viewing window.

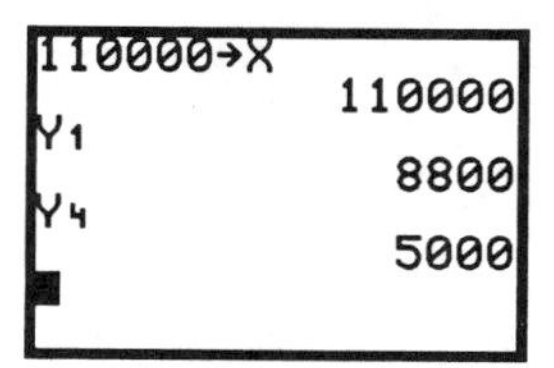

Figure 3. 27

Press [WINDOW] [▽] ; set as in Figure 3.28.

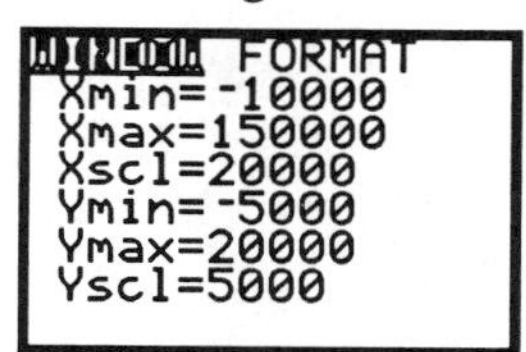

Figure 3. 28

Press [GRAPH] . See Figure 3.29.

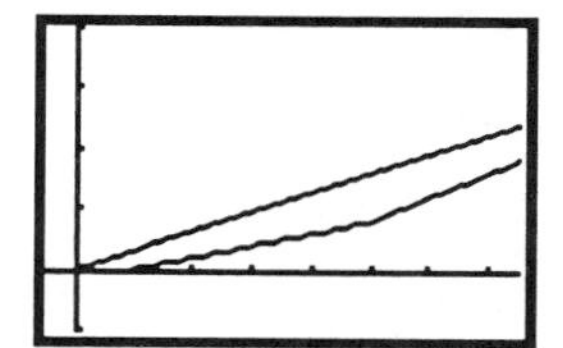

Figure 3. 29

3.2.4 Find the Intersection Point

To see the point of intersection change your WINDOW to Xmax = 350000 and Ymax = 30000. Press [GRAPH] . See Figure 3.30.

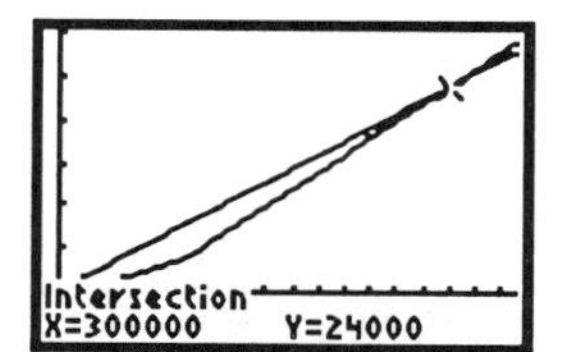

Figure 3. 30

Use [2nd] [CALC] [5:intersect] to find the point of intersection. (Refer back to section 3.1.1 if needed). You will need to select Y1 and Y4 as the pieces of the graphs that intersect . See Figure 3.30 for the coordinates of the point. This means that under the flat tax system people earning less than $300,000 per year pay more taxes than under the graduated tax plan. The graduated tax is a better system for this group of tax payers.

Chapter 4
Scientific Notation, Exponents and Logarithms

4.1 Scientific Notation

Scientific notation expresses large numbers and small numbers using powers of ten : 3250000000 = 3.25×10^9 and 0.00000123586 = 1.23586×10^{-6}. This can be done on the calculator using either $\boxed{10^x}$, the power of ten key or $\boxed{\text{EE}}$, the exponent of ten key.

> **Note**: Any decimal number can be written in scientific notation using the form $K \times 10^x$, where $1 \le K < 10$ and x is an integer. To change back to decimal number form: if $x > 0$ move x decimal places to the right, if $x < 0$ move x decimal places to the left.

Example 1:

Type the following numbers:

1. **$3.25 \text{ X } 10^9$**
2. **0.00000123589**

Press 3.25 $\boxed{\text{2nd}}$ $\boxed{10^x}$ 9 $\boxed{\text{ENTER}}$, then press 3.25 $\boxed{\text{2nd}}$ $\boxed{\text{EE}}$ 9 $\boxed{\text{ENTER}}$. Compare the results . See Figure 4.1 or 4.2. When .00000123589 is typed, it is changed to scientific notation.

> **Troubleshooting:**
> When $x \ge 10$ or $x \le -4$, in 10^x, the number is written in scientific notation.

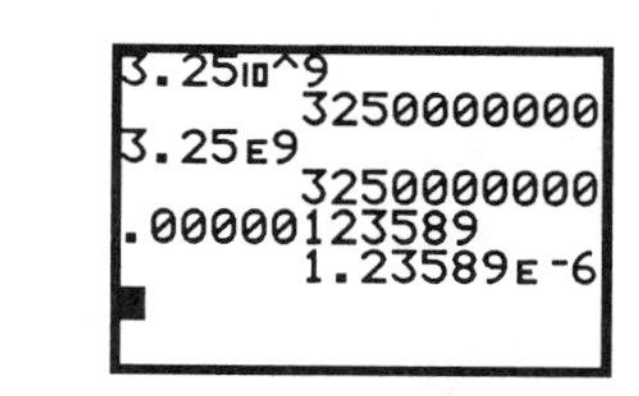

Figure 4.1 TI-82

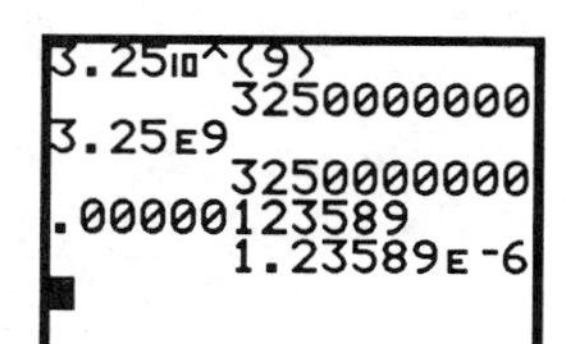

Figure 4. 2
The TI-83 will automatically insert a parenthesis when an exponent is used.

4.2 Verifying Properties of Exponents

Example 2

Verify numerically that the following are true by choosing values for *a*.

1. $a^0 = 1$
2. $a^{-1} = 1/a$
3. $a^6 = a \cdot a \cdot a \cdot a \cdot a \cdot a$

Let *a* assume various values. Enter the problems as in Figures 4.3 and 4.4. Press $\boxed{\wedge}$ for the exponent key.

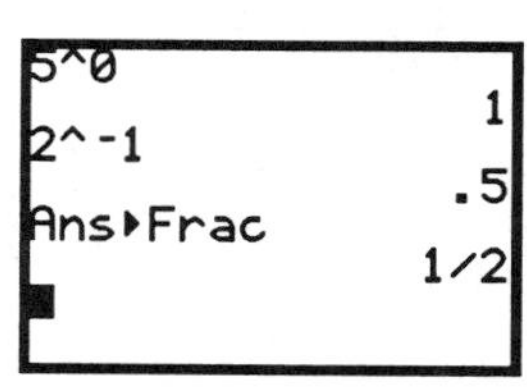

Figure 4. 3

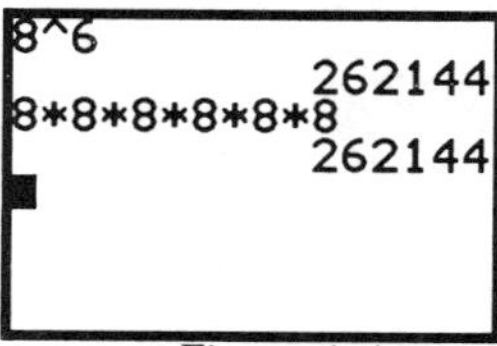

Figure 4. 4

4.2.1 Other Exponent Keys

There are other shortcut keys for exponents. The [x^2] and [x^{-1}] keys are used to paste the exponents without using the [^] key. See Figure 4.5. The cubic power and radical symbol are found under [MATH]. See Figure 4.6.

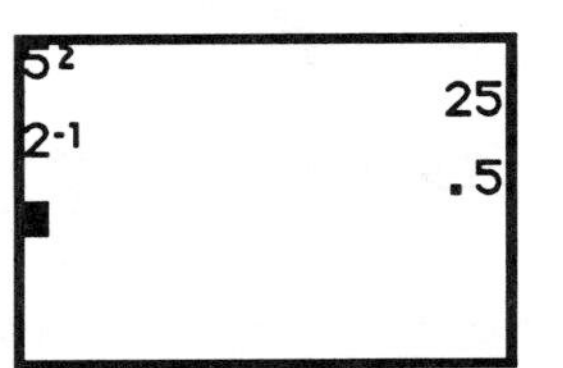

Figure 4. 5

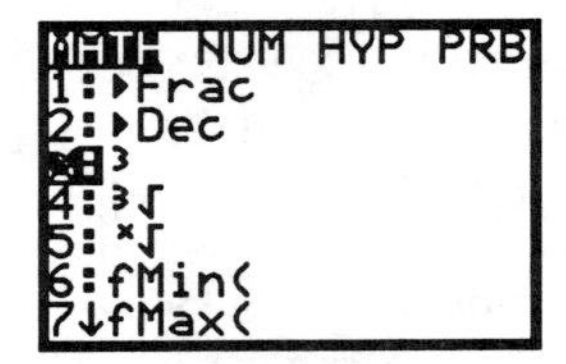

Figure 4. 6

4.2.2 Fractional Exponents

Example 3

Show that the following are equivalent

1. $25^{1/2} = \sqrt{25}$
2. $32^{1/5} = \sqrt[5]{32}$

Trouble Shooting: Fractional exponents **must** be enclosed in parentheses.

To access the square root symbol for $\sqrt{25}$, press [2nd] [√] 25 [ENTER]. See Figure 4.7.

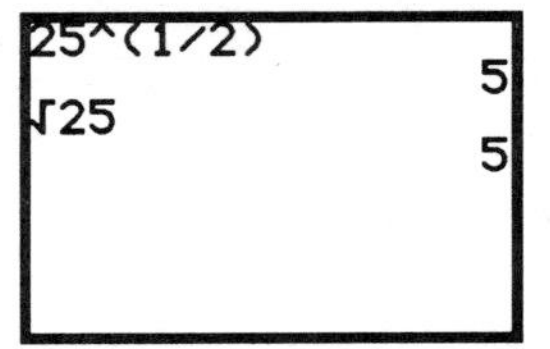

Figure 4. 7

Roots other than square root are found under the MATH menu. To type $\sqrt[5]{32}$, press 5 [MATH]; select [5: $^x\sqrt{}$] 32 [ENTER]. See Figure 4.8.

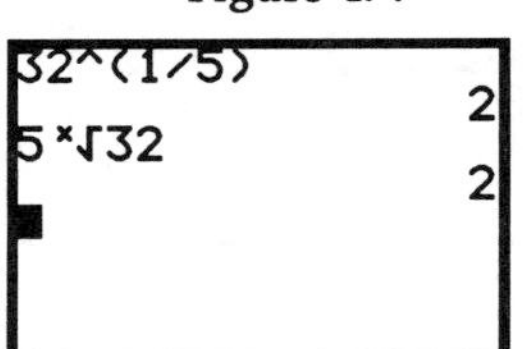

Figure 4. 8

Trouble Shooting:
The TI-82 has problems raising a negative base to a fractional power other than 1/n. For $a^{m/n}$ where $m \neq 1$, the domain is restricted to $a \geq 0$. If $a < 0$ you get an **ERR:DOMAIN** message. You have to trick the calculator into performing the operation. The TI-83 does not have this problem. For the procedure to raise -8 to the 2/3 power, see Figures 4.9 and 4.10.

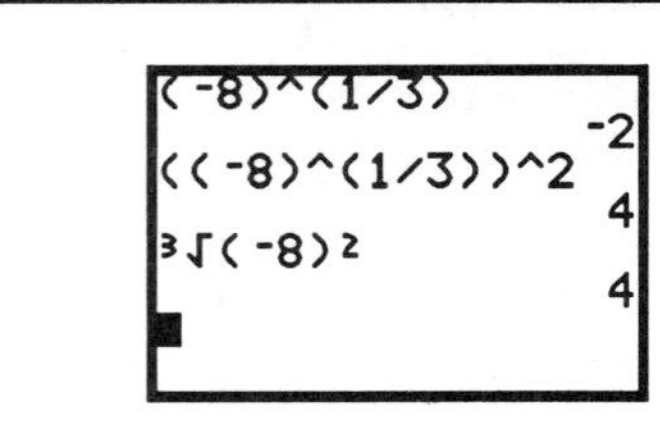

Figure 4. 9 TI-82

The TI-82 does NOT allow (-8)^(2/3), rewrite as a power to a power.

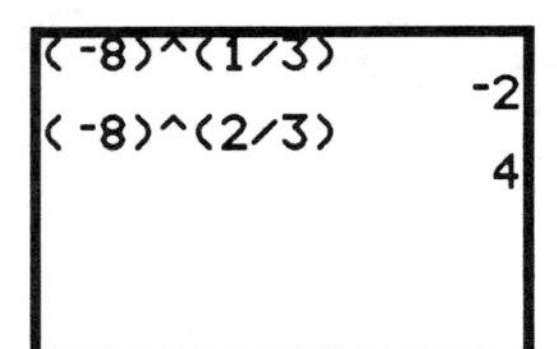

Figure 4. 10 TI-83

The TI-83 accepts (-8)^(2/3).

4.3 Using the Logarithm Key.

To find the exponent or power of ten in an equation, we use logarithms to "undo" the exponent.

If $10^x = N$ then $\log_{10} N = x$

Example 4

Write $10^x = 25$ as a logarithmic equation.

A logarithm is the value of the exponent. The solution is: $\log_{10} 25 = x$.
You read the above equation as "The logarithm of 25 to base 10 is x". To find the value of the exponent press [LOG] 25. See Figure 4.11.

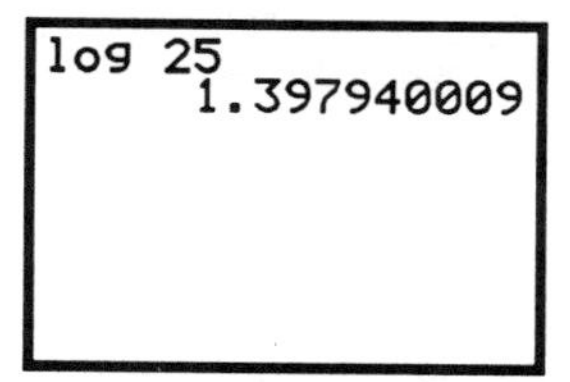

Figure 4. 11

> <u>Note</u>: $\log 25 = \log_{10} 25$. This is the common logarithm. The base 10 is understood and conventionally not written.

Check your work:

$10^{1.397940009} = 25$

Press 10 [^] [2nd] [ANS] [ENTER]

See Figure 4.12.

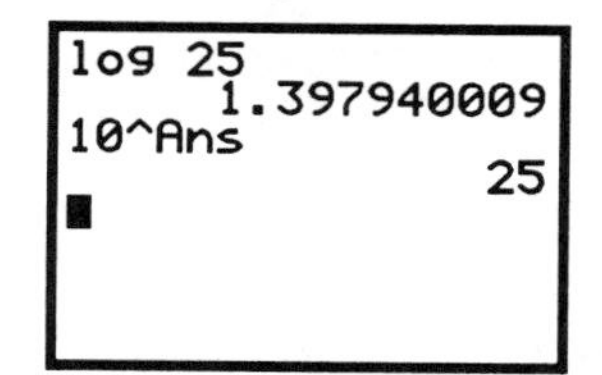

Figure 4. 12

Example 5

Sketch a graph of $y = \log x$ and use it to determine the domain and range of the function.

Press [Y=] [CLEAR] [LOG] [X,T,Θ]. See Figure 4.13. Graph the function. Press [ZOOM]; select [4:Zdecimal].

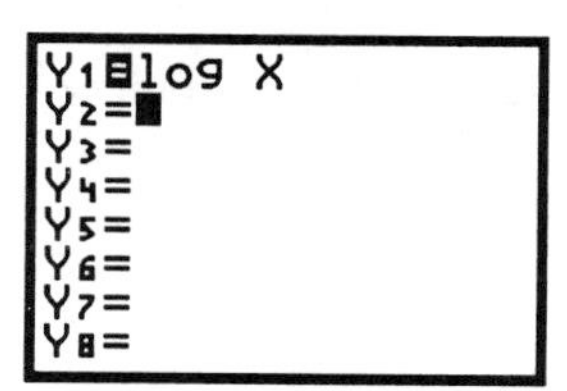

Figure 4. 13

Press [TRACE]. The graph in Figure 4.14 shows that the $\log x$ is undefined when $x = 0$. See Figure 4.14. Use [◁] to confirm that $\log x$ is undefined for $x \leq 0$. See Figure 4.15. The domain of $y = \log x$ is the set of all x such that $x > 0$.

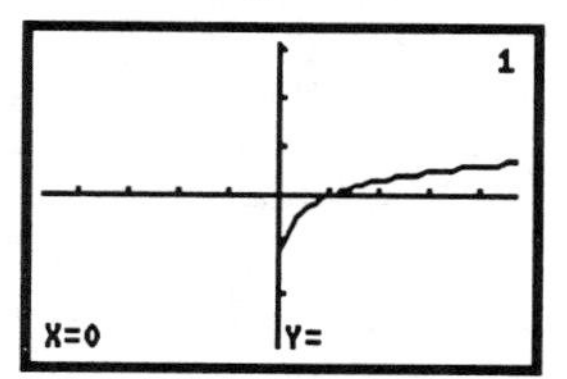

Figure 4. 14

The range is evident by looking at the graph also. As x increases y increases, but what happens as x approaches zero? y seems to be headed in a negative direction. The range for $y = \log x$ is the set of all y such that y: $(-\infty, +\infty)$, or y is any real number.

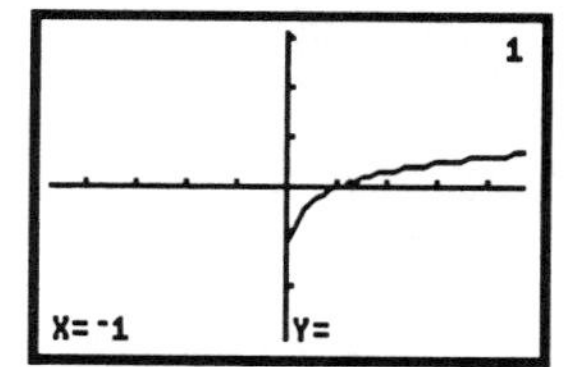

Figure 4. 15

Verify the range values by using [2nd] [TblSet] See Figure 4.16. Use [2nd] [TABLE] to see the values for $y = \log x$ for $0 < x < 1$, y is getting more negative. See Figure 4.17.

Figure 4. 16

X	Y1
0	ERROR
.01	-2
.02	-1.699
.03	-1.523
.04	-1.398
.05	-1.301
.06	-1.222

Y1=ERROR

Figure 4. 17

4.4 Solving Equations Graphically

We saw in Chapter 3 that the point of intersection represented the solution to an equation. You can solve an equation graphically by locating the point of intersection.

Example 6

Solve the equation $10^x = 25$ graphically.

Enter the following into [Y=] :

$Y1 = 10^x$

$Y2 = 25$

See Figure 4.18.

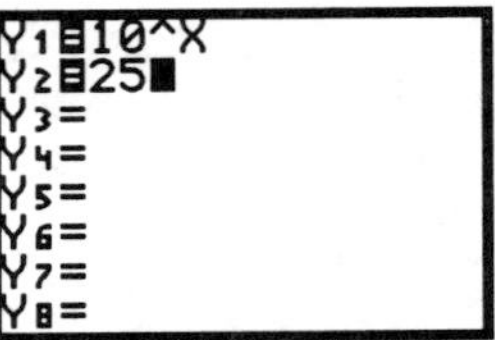

Figure 4. 18

Set the WINDOW so that both equations can be seen. See Figure 4.19.

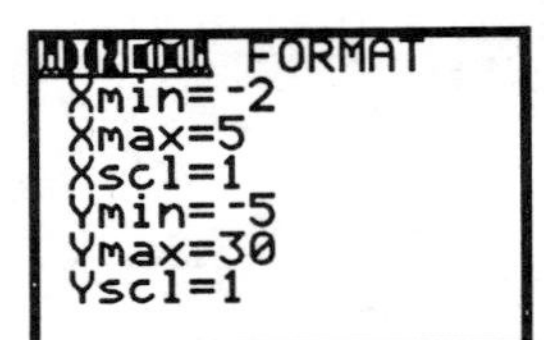

Figure 4. 19

Press [GRAPH] . See Figure 4.20.

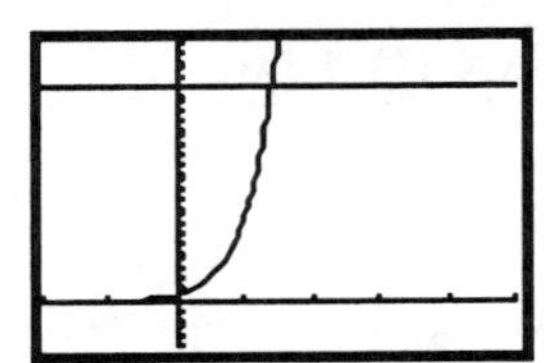
Figure 4. 20

The point of intersection represents the solution to the equation. Press [2nd] [CALC] ; select [5:intersect] . See Figure 4.21. Follow the prompts by pressing [ENTER] .

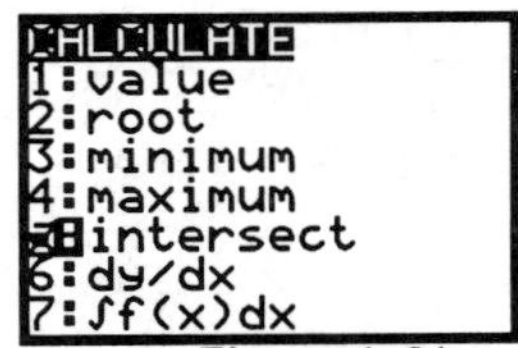

Figure 4. 21

The point of intersection occurs at approximately $x = 1.39794$. See Figure 4.22.

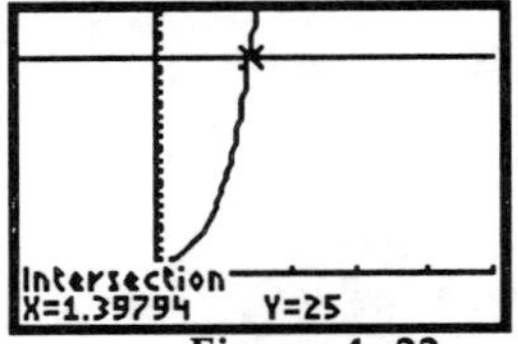

Figure 4. 22

> **Note**: The calculator remembers the intersection value for x. Immediately go to the Home Screen. Press [2nd] [QUIT] ; press [X,T,Θ] .See Figure 4.23 below.

Verify the solution by typing the expression 10^x . See Figure 4.23 below.

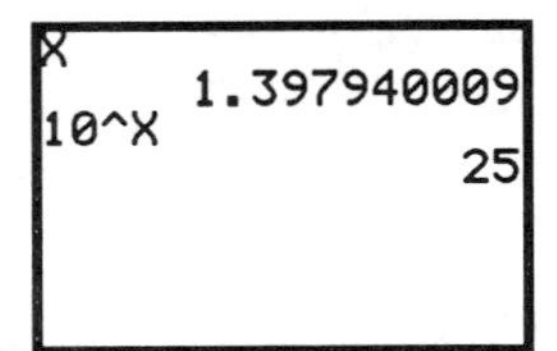

Figure 4. 23

Chapter 5
Exponential Functions

5.1 Exponential Functions

Exponential functions are in the form $y = b^x$ where $b > 0$, $b \neq 1$ and the power x can be any real number.

Example 1

Graph the following in Y1, Y2, Y3, and Y4:

1. $f(x) = 2^x$ **3.** $h(x) = 7^x$
2. $g(x) = 5^x$ **4.** $k(x) = 10^x$

Type the functions into [Y=]. See Figure 5.1. Set your [WINDOW] as in Figure 5.2. Press [GRAPH]. See Figure 5.3.

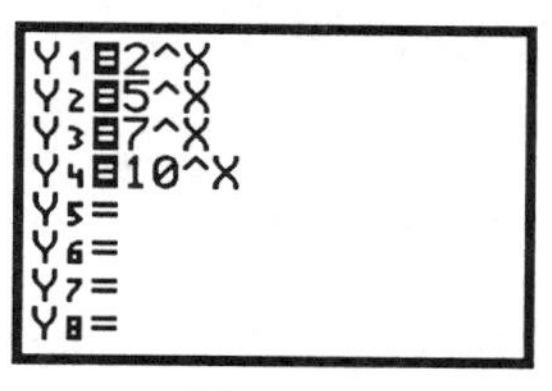

Figure 5. 1

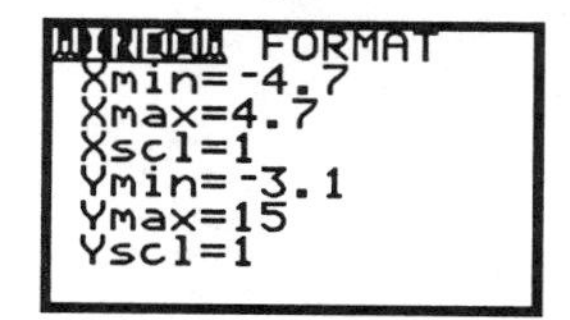

Figure 5. 2

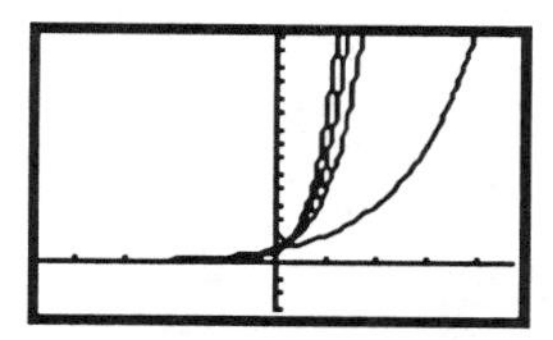

Figure 5. 3

5.1.1 Exponential Growth

All of these graphs cross the y-axis at (0, 1), because $b^0 = 1$. These exponential graphs have the same general shape. As x increases y increases slowly at first and then y increases very rapidly. The value of y is increasing at an increasing rate. This type of change is commonly called *exponential growth*.

Set up a table of values. See Figures 5.4 and 5.5. The table gives an idea of the growth rate of $y = 2^x$. As x changes from 0 to 10, Y1 changes from 1 to about 1000. But as x changes from 10 to 20, Y1 does not increase to 2000, instead Y1 now changes from about 1000 to about 1,000,000! This is a magnitude of 3 (3 powers of 10) or 10^3 times as large.

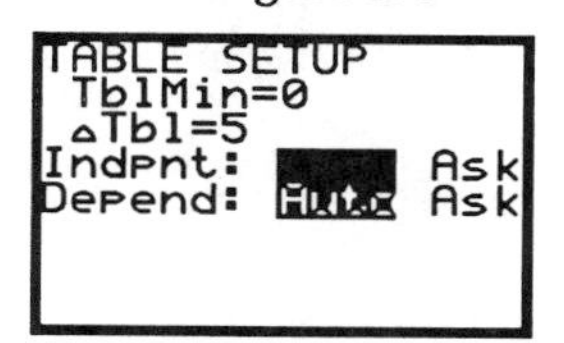

Figure 5. 4

X	Y1	Y2
0	1	1
5	32	3125
10	1024	9.77E6
15	32768	3.1E10
20	1.05E6	9.5E13
25	3.36E7	3E17
30	1.07E9	9.3E20

X=20

Figure 5. 5

5.1.2 Exponential Decay

Example 2

Graph the following Y1, Y2, Y3 and Y4:

1. $f(x) = 2^{-x}$ **3.** $h(x) = 7^{-x}$
2. $g(x) = 5^{-x}$ **4.** $k(x) = 10^{-x}$

Enter the functions as in Figure 5.6. These exponential functions differ from those in Example 1. A negative now appears in front of the exponent x.

Recall $b^{-x} = 1/b^x$, the reciprocal function. See Figure 5.7. The graphs of Figure 5.3 are reversed. Now as x increases y decreases very rapidly. This is commonly called *exponential decay*. Notice that as x gets larger y gets smaller and gets very close to zero. All of these exponential decay graphs have (0,1) as the y-intercept.

Figure 5. 6

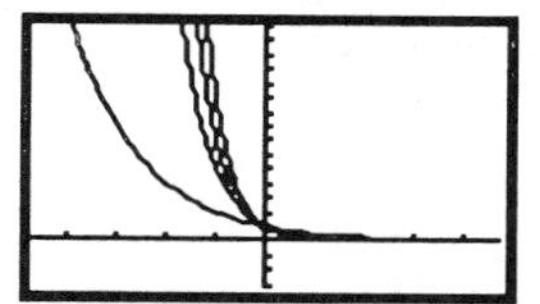

Figure 5. 7

Example 3

The population of a town with a current population of 1100 people is expected to grow at a rate of 5% per year. In how many years will the population double?

When the growth rate is a percent increase, this represents exponential growth. The function is : $P(x) = P_0 (1 + r)^x$. In this case the initial population is P_0 = 1100 and the rate of growth is r = .05.

Type the function $P(x) = 1100(1 + 0.05)^x$ in [Y=]. See Figure 5.8.

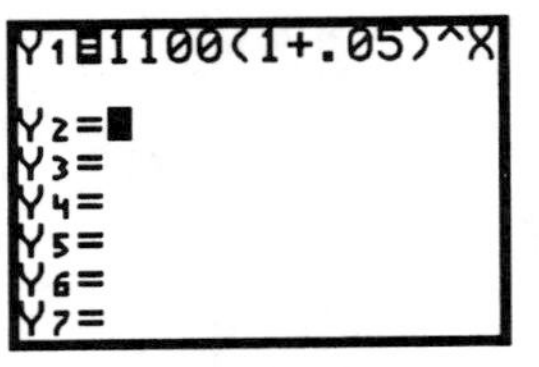

Figure 5. 8

Create a table of values. Press [2nd] [TblSet], begin with year 0 and set the increment to 1 year (Δtbl=1). See Figure 5.9.

Figure 5. 9

Press [2nd] [TABLE]. See Figure 5.10.

X	Y1
0	1100
1	1155
2	1212.8
3	1273.4
4	1337.1
5	1403.9
6	1474.1

X=0

Figure 5. 10

Approximate the doubling time by using the table. Press [▽] until y is approximately 2200, since $2P_0 = 2*1100 = 2200$. When $14 < x < 15$, y will be 2200 . See Figure 5.11). This means that the population will double in a little more than 14 years.

X	Y1
11	1881.4
12	1975.4
13	2074.2
14	2177.9
15	2286.8
16	2401.2
17	2521.2

X=15

Figure 5. 11

Example 4

Solve the above problem graphically.

Let $P(x)$= 2200.

Find the solution to:

$$2200 = 1100(1 + 0.05)^x$$

Press [Y=]. Let Y2=2200 . See Figure 5.12.

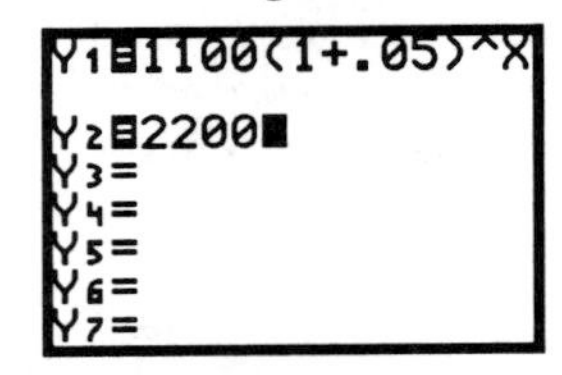

Figure 5. 12

Adjust the Window based upon the table of values. x: [-10,20] and y: [-1000,2500].

Press [GRAPH]. Find the point of intersection using the [2nd] [CALC] [5:intersect] menu . See Figure 5.13.

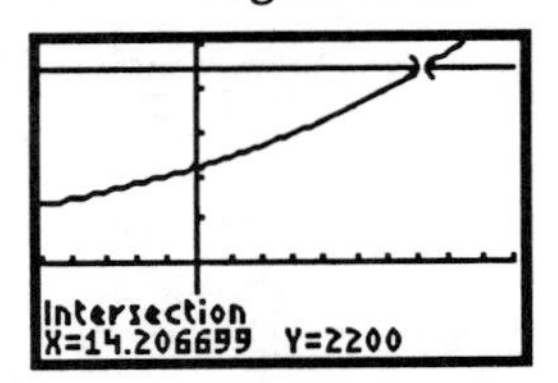

Figure 5. 13

Look at the graph of Y1 in Figure 5.13. The beginning population of 1100 is the y-intercept of the graph. The point of intersection represents the number of years it will take for the population to double, or y , to become 2200. This will occur when x is about 14.2 years.

Example 5

Graph the number of E. coli bacteria using a semi-log scale, if the bacteria grow according to the table below, where *t* is measured in 20 minute time periods.

time periods	number of E. coli
0	200
1	400
2	800
3	1,600
4	3,200
5	6,400
6	12,800
7	25,600
8	51,200
9	102,400
10	204,800

Table 5. 1

L1	L2	L3
0	200	
1	400	
2	800	
3	1600	
4	3200	
5	6400	
6	12800	

L3(1)=

Figure 5. 14

L1	L2	L3
0	200	------
1	400	
2	800	
3	1600	
4	3200	
5	6400	
6	12800	

L3=log L2

Figure 5. 15

L1	L2	L3
0	200	2.301
1	400	2.6021
2	800	2.9031
3	1600	3.2041
4	3200	3.5051
5	6400	3.8062
6	12800	4.1072

L3={2.301029995...

Figure 5. 16

1. Use the statistics menu to graph the points from Table 5.1. Press [STAT]; select [1:Edit]; enter the time periods in L1 and enter the number of bacteria in L2 . See Figure 5.14.
2. Find the log of list 2 and store to list 3. Move [Δ] to L3 label. Type [LOG] [2nd] [L2] [ENTER] . See Figures 5.15 and 5.16.
3. Set up the plot. Press [2nd] [STATPLOT] [ENTER] . Set up the plot as in Figure 5.17.
4. [CLEAR] [Y=] . Press [ZOOM] select [9:ZoomStat] . See Figure 5.18.

Figure 5. 17

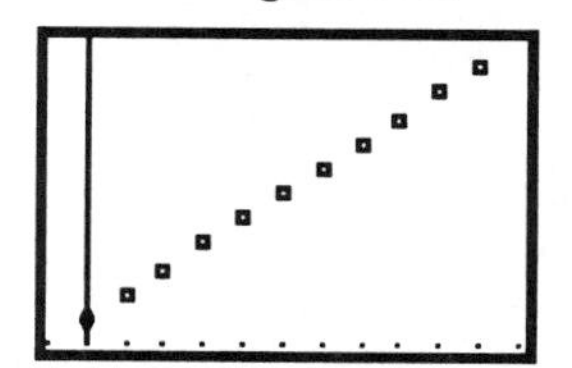

Figure 5. 18

From the data it appears that our E. coli bacteria are growing exponentially because the number is doubling every 20 minutes.

Since the number of bacteria is doubling (adding to itself), the growth rate of 100% represents a constant growth factor of 2. Therefore the plot of x against the $\log y$ forms a straight line. So $P = 200(1 + 1)^x = P=200(2)^x$.
(See Chapter 6 for more detail).

Note: Exponential data will appear linear when graphed on a semi-log scale

Example 6

Find the Linear regression equation for Example 5 , then give the algebraic explanation.
(You may have to refer to Chapter 6 for the property of logarithms to understand this example.)

EDIT CALC
1:1-Var Stats
2:2-Var Stats
3:SetUp...
4:Med-Med
5:LinReg(ax+b)
6:QuadReg
7↓CubicReg

Figure 5. 19

Find the linear regression equation using the data in L1 and L3. From the Home Screen press [STAT] <CALC>, select [:LinReg(ax+b)] . See Figure 5.19.

Press [2nd] [L1] [,] [2nd] [L3] [ENTER] . See Figures 5.20 and 5.21. The equation is:

$y = .301x + 2.301$

Figure 5. 20

LinReg
y=ax+b
a=.3010299957
b=2.301029996
r=1

Figure 5. 21

The algebraic solution will tie the linear regression equation to the exponential function. Use properties of logarithms to simplify:

$$P = 200(2)^x$$
$$\log P = \log (200(2)^x)$$
$$\log P = \log 2^x + \log 200$$
$$\log P = (\log 2)x + \log 200$$

The final equation is in the linear form

$y = mx + b.$

$m = \log 2 = .3010299957...$

$b = \log 200 = 2.30129996...$

These are the same values as ***a*** and ***b*** in the regression equation . See Figure 5.22. Therefore x is the independent value and the $\log P = y$, the dependent value.

We find the equivalent antilog values are: $10^{\log 2} = 2$ and $10^{\log 200} = 200$. See Figure 5.23. Therefore the slope of the linear regression equation allows you to find the base of the exponential function and the y-intercept allows you to find the initial value of the exponential function.

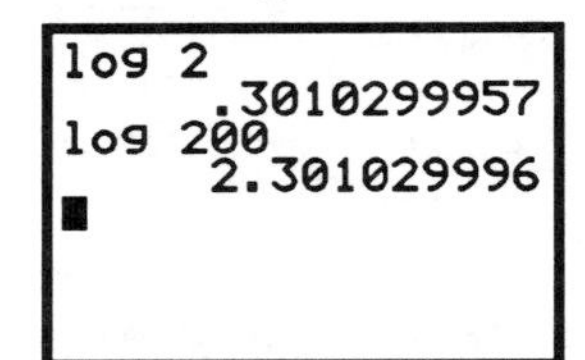

Figure 5. 22

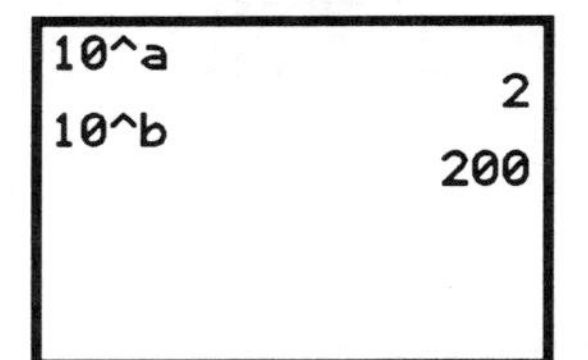

Figure 5. 23

<u>A Special Note to TI-83 Users:</u>

The TI-83 has an addition regression equation that may be helpful in this section. If your data appears to grow exponentially in the beginning, but levels off to its "carrying capacity", a logistic regression equation might be a good choice. If this kind of data is in L1 and L2, to calculate the logistic regression equation:

Press [STAT] <CALC> [:Logistic] [2nd] [L1] [,] [2nd] [L2] [ENTER] .

Chapter 6
Logarithmic and Exponential Functions

6.1 The Definition of *e*

Example 1

Consider the exponential function

$$f(x) = \left(1 + \frac{1}{x}\right)^x$$

What happens to *f(x)* as *x* gets very large?

Type the function into [Y=]. See Figure 6.1.

Figure 6. 1

6.1.1 Use TABLE to Find values

Press [2nd] [TblSet]. Put the independent variable in ASK mode. See Figure 6.2.

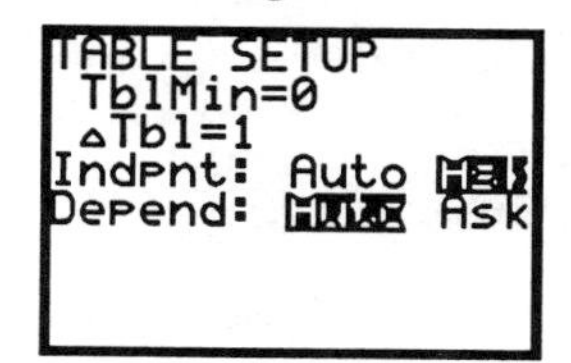

Figure 6. 2

Press [2nd] [TABLE]. Type 100 [ENTER]. See Figure 6.3. The calculator will only find the values of Y1 as you enter *x* values. Continue to enter powers of 10 values as in Figure 6.4.

X	Y1
100	2.7048

X=

Figure 6. 3

X	Y1
100	2.7048
10000	2.7181
100000	2.7183
1E6	2.7183
1E7	2.7183
1E8	2.7183
1E9	[illegible]

Y1=2.7182818271

Figure 6. 4

Interpretation:

As $x \to +\infty$, $f(x) \to 2.718\,281\,8271...$ We say that the function approaches a ***limit***, or gets close to a number value.

Example 2

Compare the limit value of *f(x)* to the calculator value of *e*.

Press [2nd] [QUIT] [CLEAR]. Store 10^9 to *x*.

Find the variable Y1 using [VARS] and e^1, press [2nd] [e^x]. See Figure 6.5.

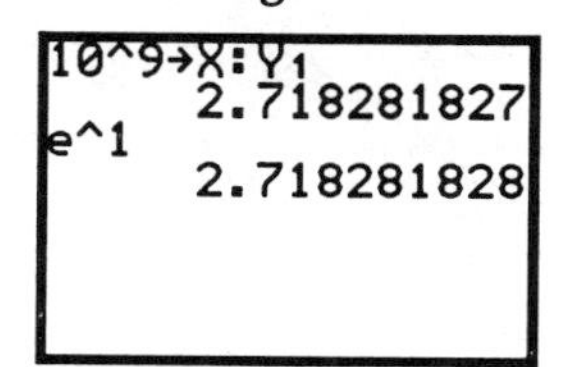

Figure 6. 5

Use the : to type two commands.

> <u>Note</u>: *e* is located above [LN] and you must supply a power.

Notice the values are nearly the same. The value of *e* is an irrational number or

$$lim_{x\to\infty}\left(1 + \frac{1}{x}\right)^x = e \approx 2.781828...$$

6.2 Draw the Inverse

Graphically speaking the inverse of a function is a reflection about the line $y = x$. The effect is that any point (a, b) becomes (b, a) on the inverse graph.

Example 3

Graph $f(x) = e^x$ and its inverse, $f^{-1}(x)$.

Press [Y=] enter e^x into Y1. Press [ZOOM] [4:ZDecimal].

To draw the inverse:

1. Press [2nd] [DRAW] ; select [8:DrawInv] . See Figures 6.6 and 6.7.
2. Type the function e^x . Press [ENTER] . After pressing ENTER, the calculator automatically goes to the graph screen. See Figure 6.8.

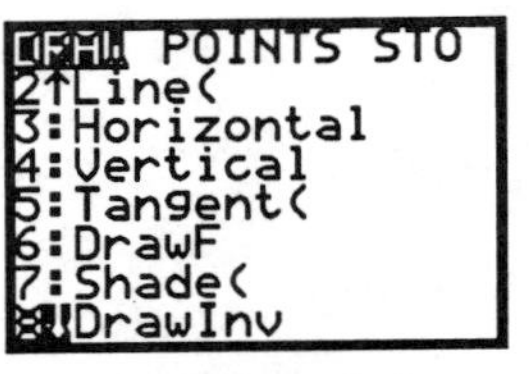

Figure 6. 6

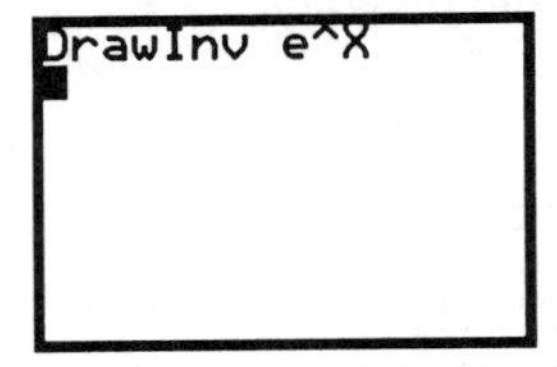

Figure 6. 7

> **Trouble Shooting:** Since you are in the draw mode you **can not** TRACE on the inverse function. To CLEAR the drawing press [2nd] [DRAW] select [1:ClrDraw].

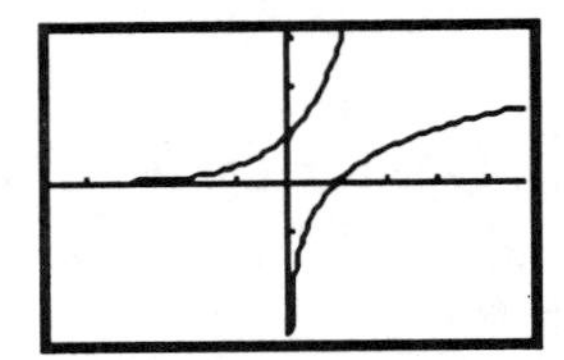

Figure 6. 8

6.1.3 Algebraically Find the Inverse Function and Graph $f^{-1}(x)$.

Example 4

Find $f^{-1}(x)$ for $y = e^x$ and graph $f^{-1}(x)$.

1. Use the definition of logarithm to "undo" the exponential function.

   if $\quad y = e^x$

   then $\quad \log_e y = x$
2. Reverse x and y to graph the function on the x, y coordinate plane:

   $y = \log_e x = \ln x$

   Using function notation:

   if $\quad f(x) = e^x$

   then $\quad f^{-1}(x) = \ln x.$
3. Graph $y = \ln x$.

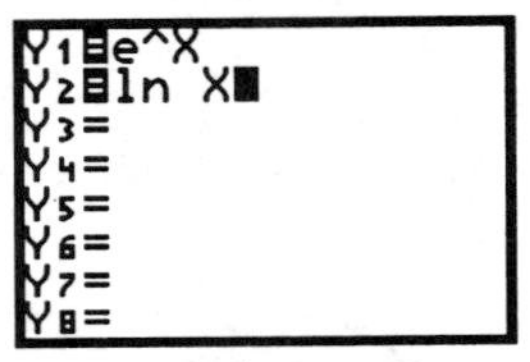

Figure 6. 9

Press [Y=] into Y2; type [LN] [X,T,Θ] . See Figure 6.9. Press [GRAPH] . See Figure 6.10.

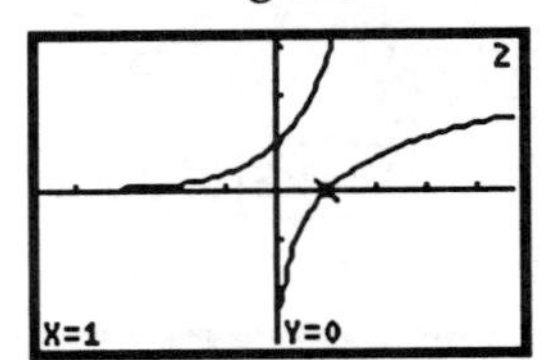

Figure 6. 10

> **Note**: Now you can TRACE on the inverse graph. The point (1, 0) on graph Y1 is the point (0, 1) on the inverse graph, Y2.

6.2 Graphing Common and Natural Logarithms

The graphing calculator has two built-in logarithms:

1. For the common logarithm, $\log_{10} x$, use the [LOG] key.
2. For the natural logarithm, $\log_e x$ or $\ln x$, use the [LN] key.

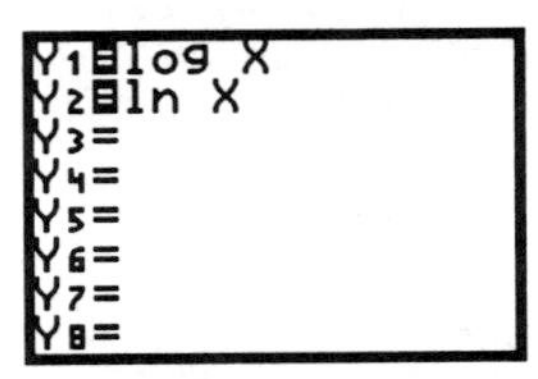

Figure 6. 11

Example 5

Compare the graphs of :

$f(x) = \log x$ and $g(x) = \ln x$.

Enter the functions into Y1 and Y2 then select the window with x:[0,4.7] and y: [-4,4]. See Figures 6.11 and 6.12.

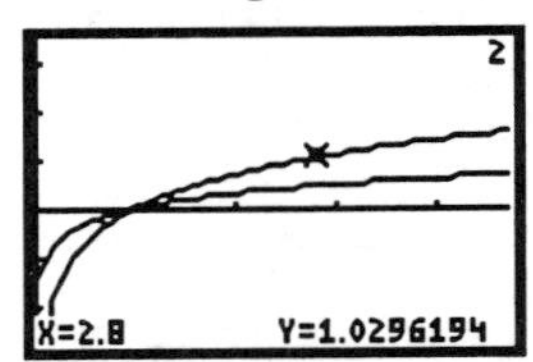

Figure 6. 12

Interpretation

Since the logarithm is the inverse of the an exponential function, the value of the logarithm represents the exponent of an exponential function. Thus, the larger the base of the logarithm the slower the exponent grows. The function *ln x* grows faster than *log x*. The graph of *ln x* is on top of *log x* for $x > 1$.

6.2.1 Graphs of Logarithms to Bases Other Than 10 and *e*.

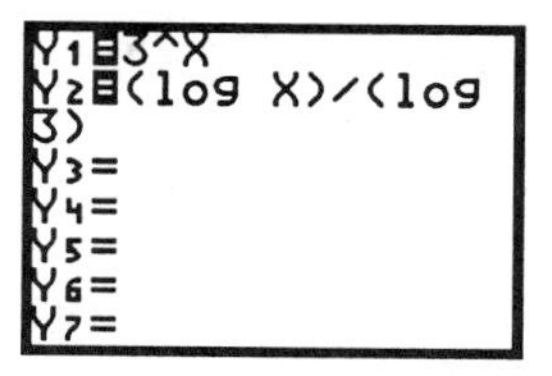

Figure 6. 13

Example 6

Find the inverse of $f(x) = 3^x$ and graph it.

1. By definition $f^{-1}(x) = \log_3 x$, which is not a built-in function on the calculator.
2. Use Algebra and properties of logarithms to find the inverse function.

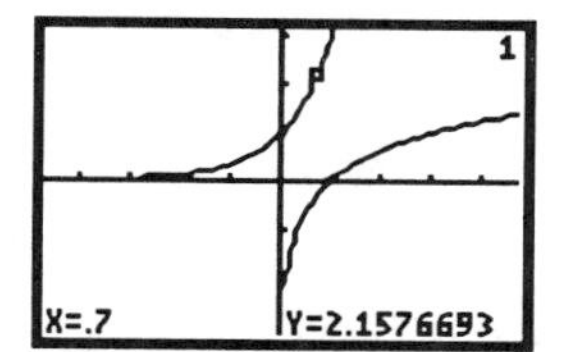

Figure 6. 14

$y = 3^x$		
$x = 3^y$	exchange x and y	
$\log x = \log 3^y$	take the *log* of both sides	
$\log x = y \log 3$	exponent property	
$\log x/\log 3 = y$	divide by $\log 3$	
$y = \log x/\log 3$		

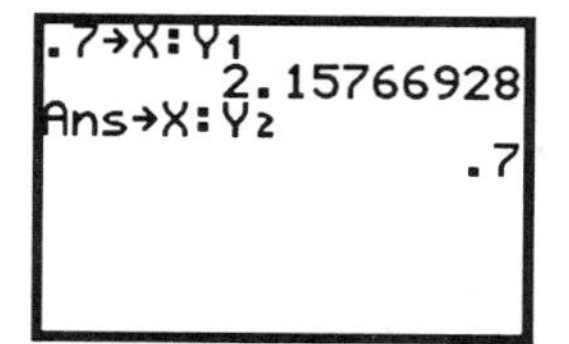

Figure 6. 15

> **Note**: The change of base formula can be used, $\log_b N = \log N/\log b = \ln N/\ln b$

3. Graph both 3^x and $\log x/\log 3$, then compare them.

   Press [Y=] enter the functions . See Figure 6.13.

   Press [ZOOM] ; select [4:Zdecimal] . See Figure 6.14. [TRACE] to $x = 0.7$

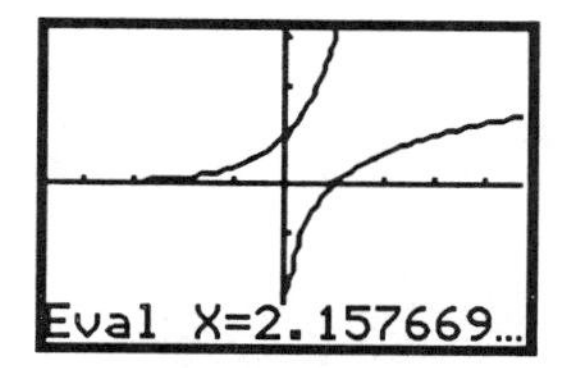

Figure 6. 16

4. Check numerically:

   $Y1 = 3^{.7} = 2.15766928$

   $Y2 = \log 2.15766928/\log 3 = .7$

   See Figure 6.15.

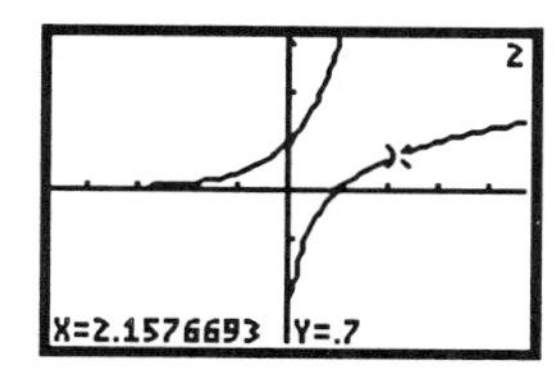

Figure 6. 17

Verify the point on the inverse graph.

Press [2nd] [CALC] ; select [1:value].

Type in 2.15766928 [ENTER] . See Figure 6.16. [▽] takes you to the point on the inverse graph. See Figure 6.17.

Chapter 7
Power Functions

7.1 Power Functions with Positive Integral Powers

A power Function has the form

$$y = kx^p$$

where k and p are constants. The simplest power function is $y = x$, where $p = 1$. Its graph is linear. However not all other power functions look linear.

7.1.1 Visualizing Odd Power Functions

Example 1

Graph the following and generalize the shape of the graph.

$Y1 = x^3$

$Y2 = x^5$

$Y3 = x^7$

Press [Y=] [CLEAR] to erase all old functions. Enter the above functions See Figure 7.1. Set your WINDOW as in Figure 7.2. Press [GRAPH] . See Figure 7.3.

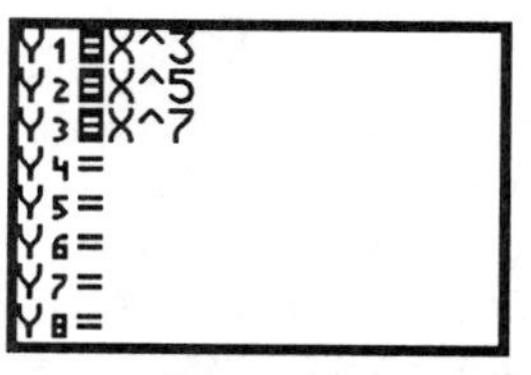

Figure 7. 1

WINDOW FORMAT
Xmin=-2
Xmax=2
Xscl=1
Ymin=-4
Ymax=4
Yscl=1

Figure 7. 2

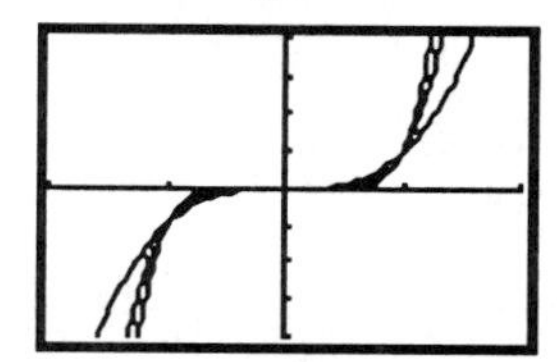

Figure 7. 3

The graphs go through the origin (0, 0), have both positive and negative y values and have a "lazy S " shape. As you move left to right the functions are always increasing. A look at the table of values, in Figure 7.4, shows that for the odd power functions as x increases to $+\infty$, y increases pretty fast to $+\infty$. As x decreases to $-\infty$ y decreases to $-\infty$.

X	Y1	Y2
-300	-2.7E7	-2E12
-200	-8E6	-3E11
-100	-1E6	-1E10
0	0	0
100	1E6	1E10
200	8E6	3.2E11
300	2.7E7	2.4E12

X=-300

Figure 7. 4

Press [ZOOM] ; select [3:Zoom Out] . See Figure 7.5. The calculator gives you a chance to reposition your cursor; then press [ENTER] . See Figure 7.6. This shows that all the graphs have a similar shape or "global behavior."

Figure 7. 5

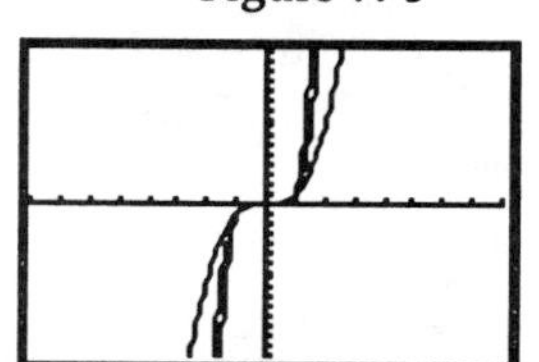

Figure 7. 6

7.1.2 Visualizing Even Power Functions

Example 2

Graph the following and generalize the shape of the graph.

$Y1 = x^2$

$Y2 = x^4$

$Y3 = x^6$

Enter the functions into [Y=]. Reset the WINDOW as in Figure 7.2 or press [ZOOM] <Memory>; select [1:ZPrevious]. Press [GRAPH]. See Figure 7.7.

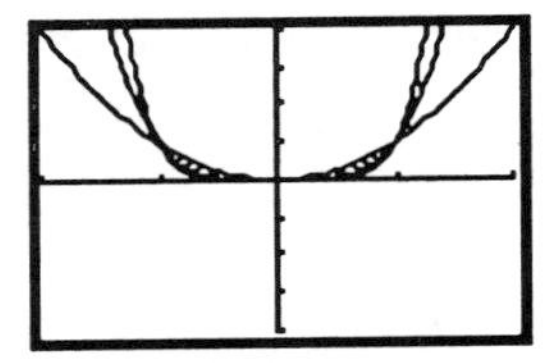

Figure 7. 7

The graphs go through the origin (0, 0), have only positive y values and have a "U" shape. As you move from negative values to positive values of x the functions have large positive values then decrease to zero then increase again. A look at the table of values, in Figure 7.8, shows that for the even power functions, as x increases to $+\infty$, y increases pretty fast to $+\infty$. As x decreases to $-\infty$., y increases to $+\infty$.

X	Y1	Y2
-300	90000	8.1E9
-200	40000	1.6E9
-100	10000	1E8
0	0	0
100	10000	1E8
200	40000	1.6E9
300	90000	8.1E9

X=-300

Figure 7. 8

7.1.3 Polynomial Functions

When positive integer power functions are added together you get a polynomial function:

$$y = a_n x^n + a_{n-1} x^{n-1} + ... + a_1 x^a + a_0$$

where a_n is a constant coefficient ($a_n \neq 0$), and n is a positive integer power.

Example 3

The deer population in a national forest was monitored over a 25 year period. The data collected can be modeled by the fifth-degree polynomial:

$$D(x) = -0.125x^5 + 3.125x^4 + 4000.$$

Graph the function and interpret the graph.

Let Y1= $-.125x^5 + 3.125x^4 + 4000$. See Figure 7.9.

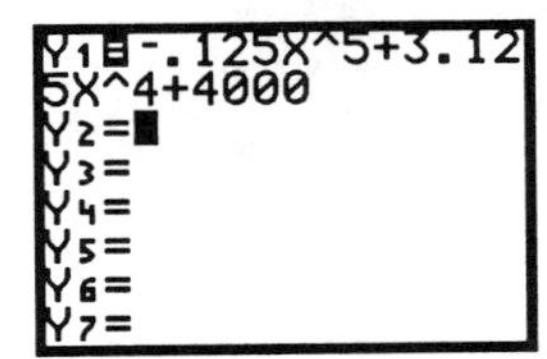

Figure 7. 9

> **Note**: Use the table of values to find the appropriate WINDOW.

Press [2nd] [TblSet]. Let TblMin = 0 and Δtbl = 1. Press [2nd] [TABLE]. See Figures 7.10 and 7.11. It looks like the maximum population occurs around 20 years.

X	Y1
0	4000
1	4003
2	4046
3	4222.8
4	4672
5	5562.5
6	7078

X=6

Figure 7. 10

X	Y1
19	101741
20	104000
21	101241
22	91846
23	73960
24	45472
25	4000

X=20

Figure 7. 11

Set the WINDOW as in Figure 7.12.

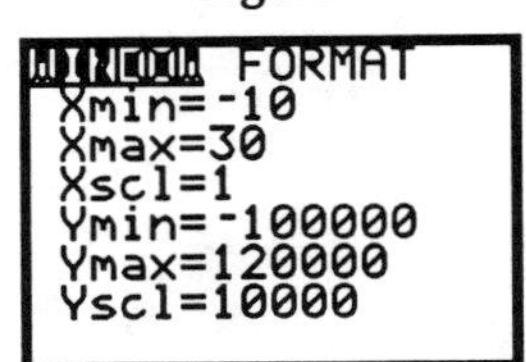

Figure 7. 12

Press [GRAPH], then [TRACE] to explore the graph. See Figure 7.13.

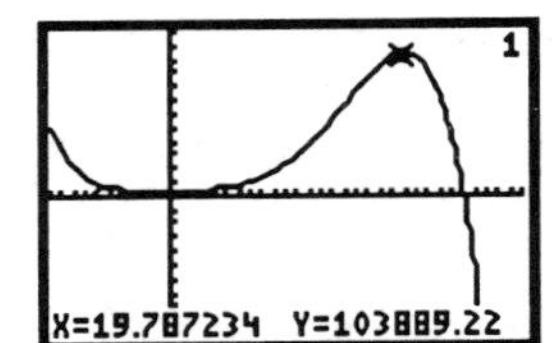

Figure 7. 13

Interpretation:

It took 20 years for the deer population to go from 4000 to a maximum population of about 104,000. Over the next five year period the population decreased back to 4000. The sharp decrease was probably a result of a lack of food supply caused by over population and/or disease.

7.1.4 Power Regression Equations.

Example 4

Below are some data reported on AIDS in women. Find a power regression equation that models the data.

Year	AIDS Cases
1	18
2	30
3	36
4	92
5	198
6	360
7	631
8	1016
9	1430

L1	L2	L3
1	18	------
2	30	
3	36	
4	92	
5	198	
6	360	
7	631	

L1(1)=1

Figure 7. 14

> Note: Clear functions from [Y=] and turn OFF all plots.

1. Enter the data in L1 and L2. Press [STAT] [1:Edit] . See Figure 7.14. Press [2nd] [STATPLOT] [1] and set up the plot as in Figure 7.15. Press [ZOOM] ; select [9:ZoomStat]. See Figures 7.16 and 7.17.
2. Find the regression equation. Press [STAT] [▷] to <CALC> ; select [B:PwrReg] [2nd] [L1] [,] [2nd] [L2] [ENTER] . See Figures 7.18 and 7.19.
3. Put the equation into Y1. To paste: *Press [Y=] [VARS] ; select[5:Statistics] [▷] [▷] to <EQ> ; select [:RegEQ]. Press [GRAPH] . See Figure 7.20.

> * TI-83 NOTE: Press [STAT] [▷] to <CALC>. Select [A:PwrReg] , [2nd] [L1] [,] [2nd] [L2] [,] [Y1] to directly store the equation into Y1

The power function fit is pretty good in the beginning, but not so good at the end.

Figure 7. 15

Figure 7. 16

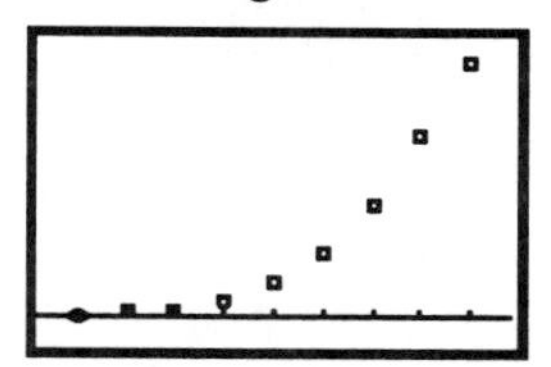

Figure 7. 17

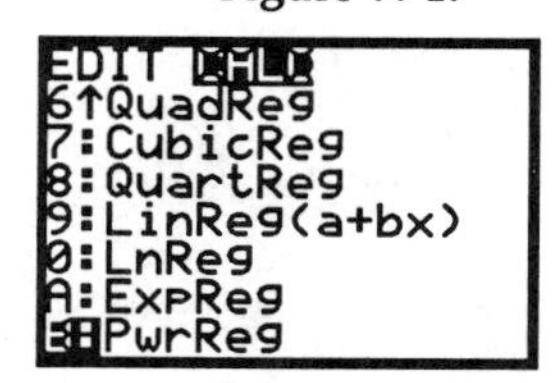

Figure 7. 18

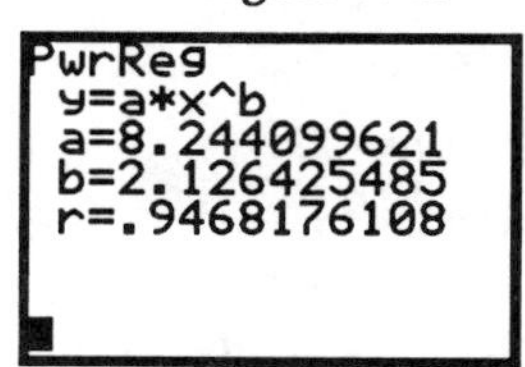

Figure 7. 19

TI-83. Turn DiagnosticsOn under CATALOG to see r.

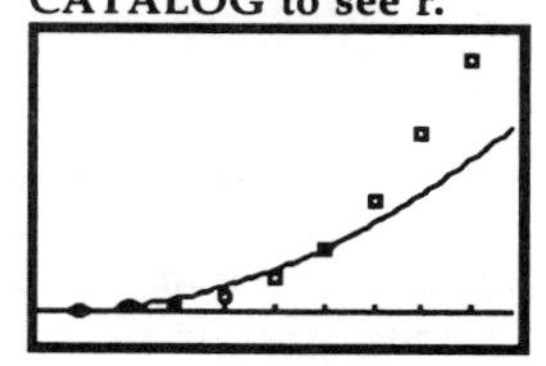

Figure 7. 20

Example 5

Find other polynomial regression equations or exponential equation .

Repeat steps 1 - 3 above choosing different regression equations. Then use these models to predict future AIDS values. Graphing screens will not be shown.

Example 6

Compare the graphs of negative integer power exponents and generalize about odd and even negative integer powers.

First Graph:

Y1 = x^{-1}

See Figure 7.21.

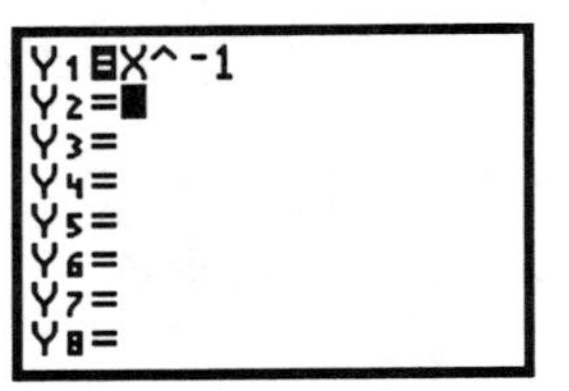

Figure 7. 21

Press [ZOOM] select [4:Decimal] . See Figure 7.22. This graph is in two pieces, since $x^{-1} = 1/x$, so $x \neq 0$, and it is symmetric to the origin.

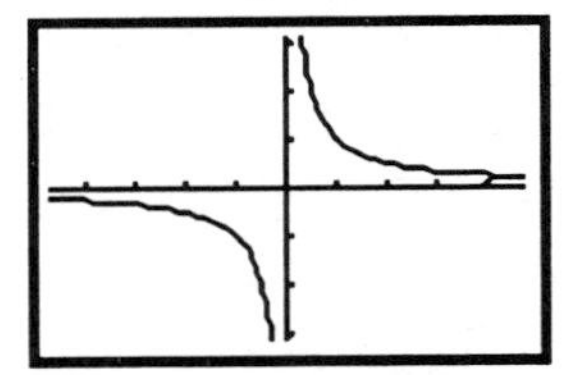

Figure 7. 22

Graph other negative odd integer powers:

Y2 = x^{-3}

Y3 = x^{-5}

Enter the functions into Y2 and Y3. See Figure 7.23. Press [GRAPH] . See Figure 7.24.

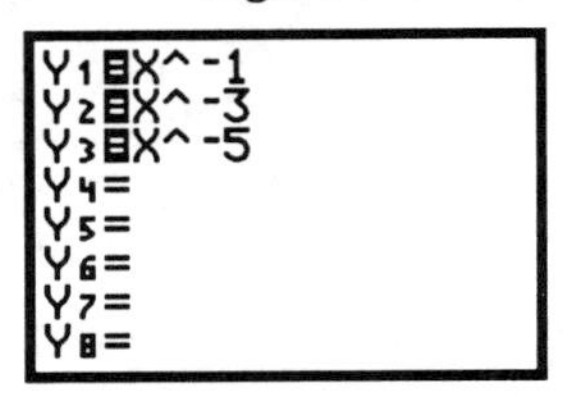

Figure 7. 23

To the right of $x = 0$, the function is at a large negative value and decreases rapidly to zero. As $x \to +\infty$, $y \to 0$

To the left of $x = 0$, the function is near zero and decrease rapidly to large negative values. As $x \to -\infty$, $y \to 0$

Figure 7. 24

Press [Y=] enter an even negative integer power function:

Y1 = x^{-2}

Press [GRAPH] . See Figures 7.25 and 7.26.

This is a reasonable graph because $x^{-2} = 1/x^2$, so $x \neq 0$ and all y values are positive. This graph is symmetric to the ***y* -axis.**

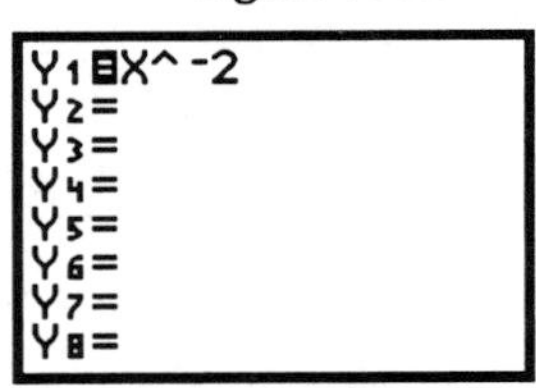

Figure 7. 25

Press [Y=] enter other even negative integer powers:

Y2 = x^{-4}

Y3 = x^{-6}

Press [GRAPH] . See Figure 7.27.

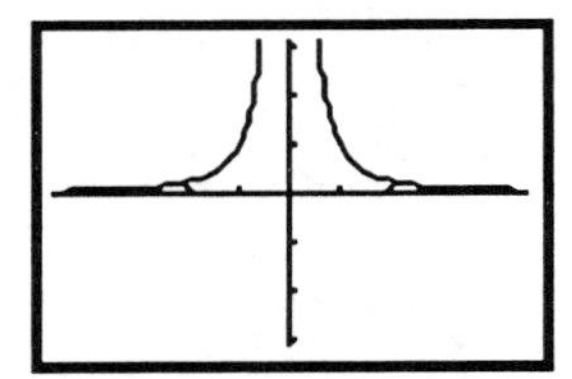

Figure 7. 26

Once again the functions have the same general shape as x^{-2} with all positive values. On either side of $x = 0$, the function goes to a positive infinity. As one moves left to right the curve begins near zero and increases to $+\infty$. At $x = 0$,, the function is undefined. To the right of zero the function is at $+\infty$ and decreases rapidly to zero.

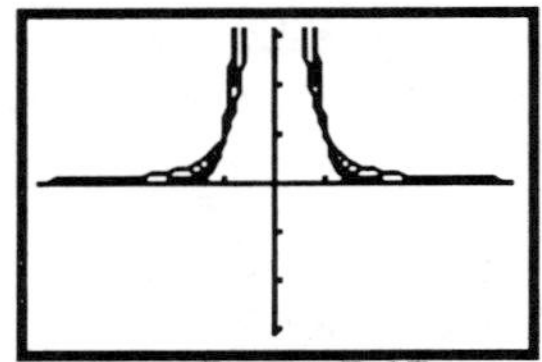

Figure 7. 27

> **Troubleshooting:** For x^n where n is a decimal value (non-integer) the domain of the function is limited to $x>0$. You will only see one-half of the power function.

7.2 Log -log Plots

When data is collected and graphed by hand, log-log paper is sometimes used. We will now explore how power functions behave on log-log plots.

Note: Refer back to Chapter 5 for semi-log plots.

Example 7

Graph the power function data and its formula:

L1	1	2	3	4	5	6	7
L2	1	8	27	64	125	216	343

1. Press STAT ; select [1:Edit]. Enter the data into L1 and L2 . See Figure 7.28. Notice that L2 = L1^3. A function that would describe this relation is $y = x^3$.
2. Into Y= type x^3 . See Figure 7.29.
3. Set up the plot . See Figure 7.30. Press ZOOM ; select [9:ZoomStat].

7.2.1 Graph the data as a log-log plot.

1. Let ***log*** L1 = L3 and ***log*** L2 = L4 . See Figure 7.31.
2. CLEAR Y=
3. Set up the plot with L3 and L4 . See Figure 7.32.
4. Press ZOOM ; select [9:ZoomStat] . See Figure 7.33.

Note: Power functions appear to be linear when graphed on a log-log plot.

Verify algebraically:

$$y = x^3$$

$$\log y = \log x^3$$

$$\log y = 3\log x$$

When you graph this equation, ***log x*** is the independent variable (input value) and ***log y*** is the dependent variable (output value). Thus we have a direct proportion with the 3 representing the slope of a line. Thus the slope will always be the power of the power function of a *log-log* plot.

Note: When data is collected, to test if the data has a power function relationship graph the *log* of list 1 with the *log* of list 2. If the graph appears to be linear, the relationship is a power function. The slope indicates the power of the function.

L1	L2	L3
1	1	------
2	8	
3	27	
4	64	
5	125	
6	216	
7	343	

L2={1,8,27,64,1...

Figure 7. 18

Y1=X^3
Y2=
Y3=
Y4=
Y5=
Y6=
Y7=
Y8=

Figure 7. 29

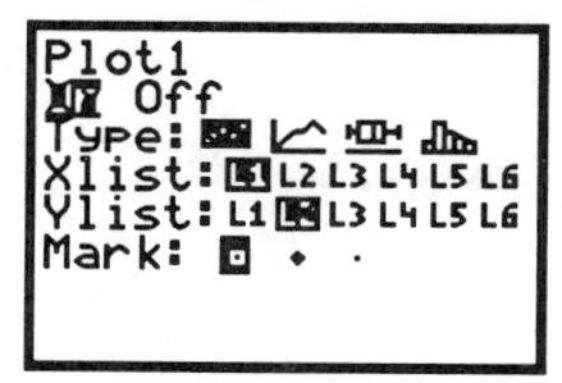

Figure 7. 30

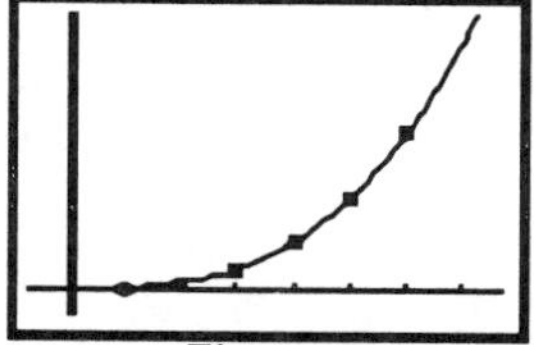

Figure 7. 31

Figure 7. 32

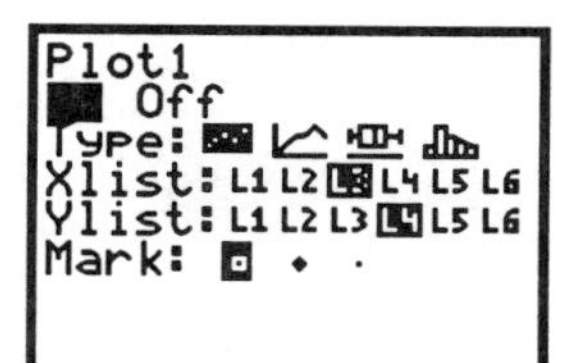

Figure 7. 33

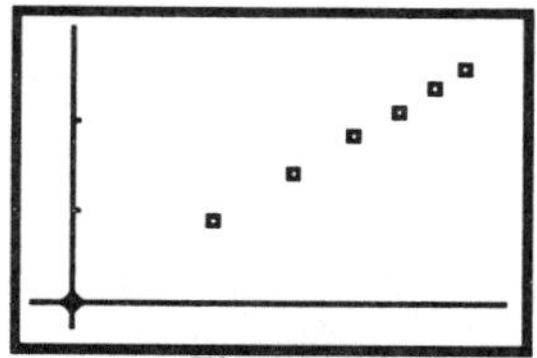

Figure 7. 4

Chapter 8
Quadratic and Polynomial Functions

8.1 Quadratic Functions

Functions of the form $y = ax^2 + bx + c$ are called quadratic or second degree equations.

> Note: The highest power of x is two.
> All quadratic equations have the shape of a parabola.

Example 1

Graph $y = x^2 - x - 6$ and identify a, b, and c.

For this quadratic equation:

$a = 1, b = -1$ and $c = -6$

To graph press [Y=] [CLEAR]. Into Y1 enter $x^2 - x - 6$. Press [ZOOM] [6], since a=1 the graph opens up. See Figure 8.1.

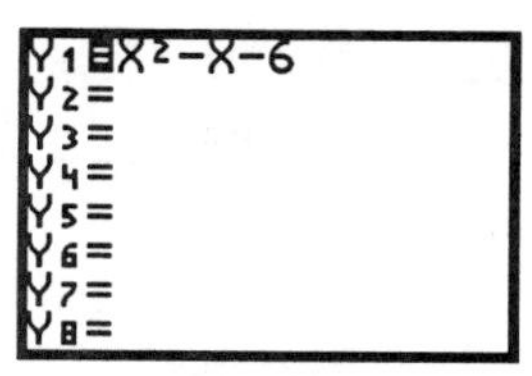

Figure 8. 1

8.1.1 Finding y-intercept.

Press [TRACE] to find the y-intercept; if $x = 0$ then $y = -6 = c$. See Figure 8.2.
The y-intercept is (0, -6) = (0, c).

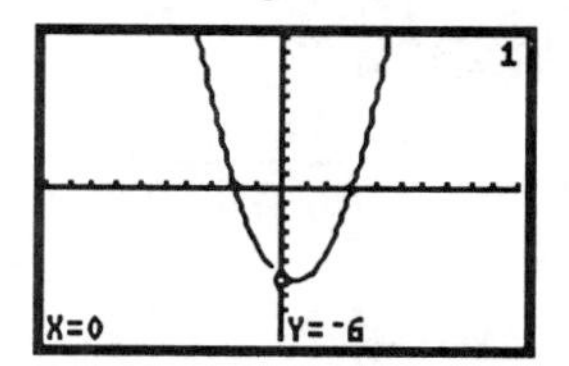

Figure 8. 2

8.1.2 Finding x-intercepts / Finding the Roots/ Finding the zeros.

Use [TRACE] and arrows to estimate where the graph crosses the x-axis. There are two ***x-intercepts*** (where $y = 0$), also known as ***roots*** or ***zeros.*** The left intercept is around $x = -2$ and the right intercept is around $x = 3$. See Figures 8.3 and 8.4.

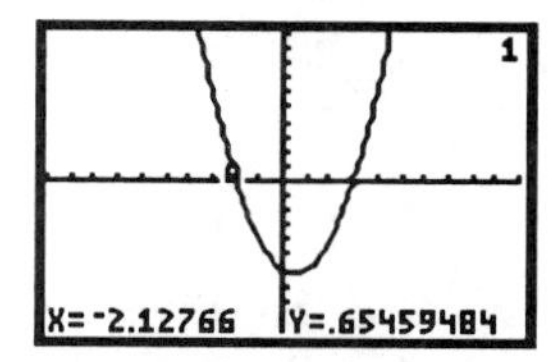

Figure 8. 3

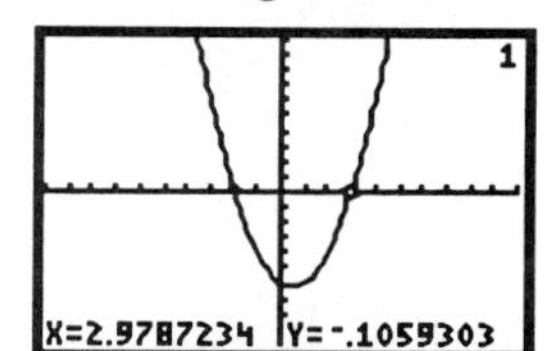

Figure 8. 4

Use the calculator to find the left x-intercept . See Figure 8.5 by following the steps below. See Figures 8.6 -8.11 for TI-82, TI-83 differences.

Press [2nd] [CALC] [2]. Follow these steps:

1. Position the cursor <u>to the left</u> of the x -intercept when prompted. Press [ENTER]
2. Reposition the cursor <u>to the right</u> of the x -intercept when prompted. Press [ENTER].
3. When prompted GUESS? position the cursor near the x -intercept. Press [ENTER]. The root or x -intercept value is displayed . See Figure 8.5).

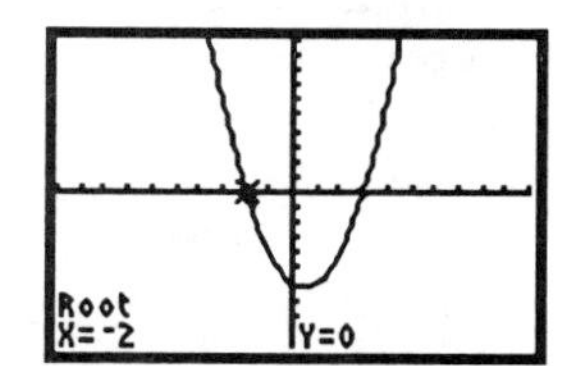

Figure 8. 5

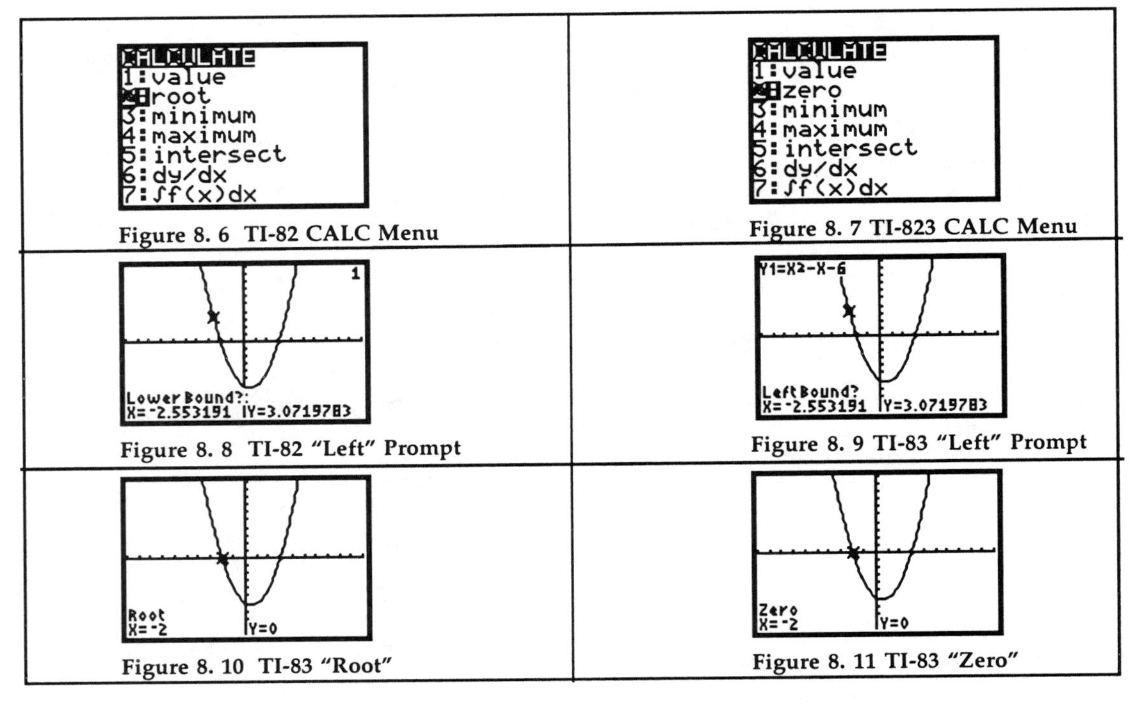

Figure 8. 6 TI-82 CALC Menu

Figure 8. 7 TI-823 CALC Menu

Figure 8. 8 TI-82 "Left" Prompt

Figure 8. 9 TI-83 "Left" Prompt

Figure 8. 10 TI-83 "Root"

Figure 8. 11 TI-83 "Zero"

Find the right x - intercept:
Repeating the process we find that the other x - intercept is (3, 0). See Figure 8.12. There are two roots, at $x = -2$ and $x = 3$.

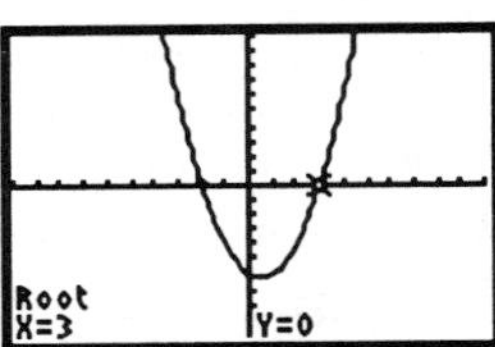

Figure 8. 12

8.1.3 Check Roots Algebraically:

$y = x^2 - x - 6$

if $x = -2$, $y = (-2)^2 - (-2) - 6 = 0$

if $x = 3$, $y = 3^2 - 3 - 6 = 0$. See Figure 8.13.

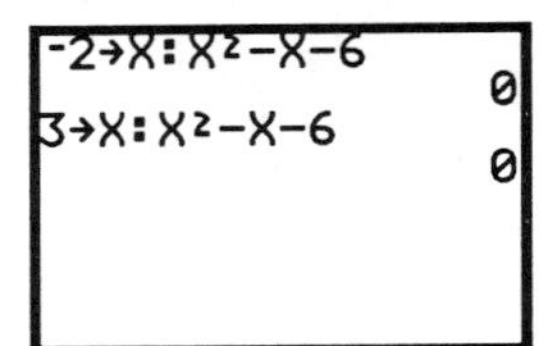

Figure 8. 13

8.1.4 Finding the Vertex of a Quadratic Function

We see from the graph in Figure 8.12 that the function $y = x^2 - x - 6$ opens upward. This will be true when $a > 0$. As we move left to right the ***minimum*** point on the graph is called the ***vertex.*** It is somewhere between $x = 0$ and $x = 2$.

Example 2

Find the vertex for $y = -2x^2 - 12x - 13$.

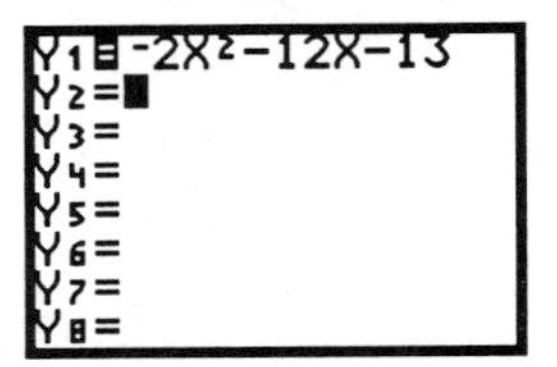

Figure 8. 14

Notice that $a < 0$ and the graph opens downward. See Figures 8.14 and 8.15. Now the vertex is the ***maximum*** point on the graph. The vertex appears to be around the point (-3, 5).

Use the calculator to find the vertex.

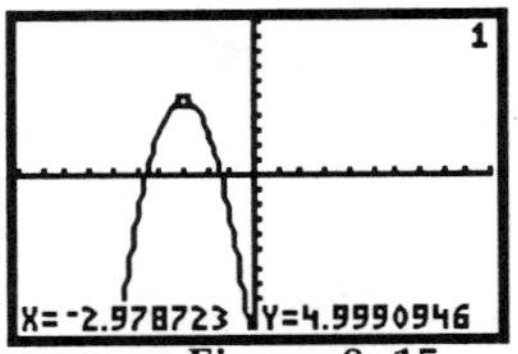

Figure 8. 15

Press [2nd] [CALC] [4] for maximum . See Figure 8.16. Then follow these steps*:

1. Position the cursor ***to the left*** of the maximum (lower bound) when prompted. Press [ENTER] . See Figure 8.17.
2. Reposition the cursor ***to the right*** of the maximum (upper bound) when prompted . See Figure 8.18. Press [ENTER]

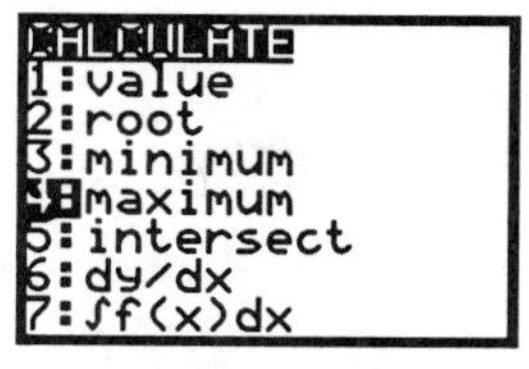

Figure 8. 16

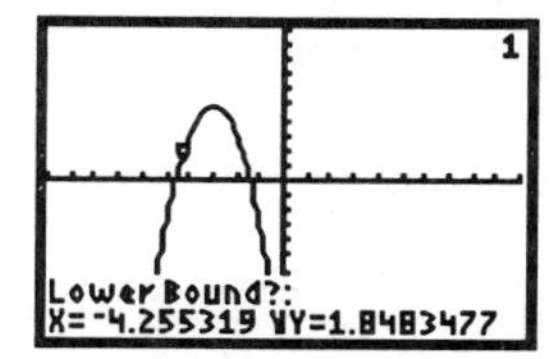

Figure 8. 17

When prompted GUESS? position the cursor near the maximum. Press [ENTER] . See Figure 8.19. The vertex is at the point (-3, 5) . See Figure 8.20.

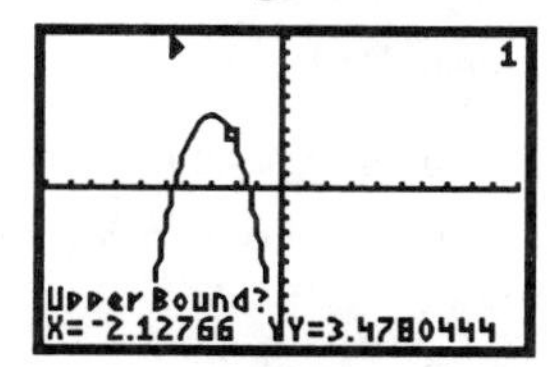

Figure 8. 18

> ***TI-83 Note**: As with the x - intercept the TI-83 prompts for left and right bounds. You can also enter a number guess for the maximum. Only the TI-82 screens are shown to the right.

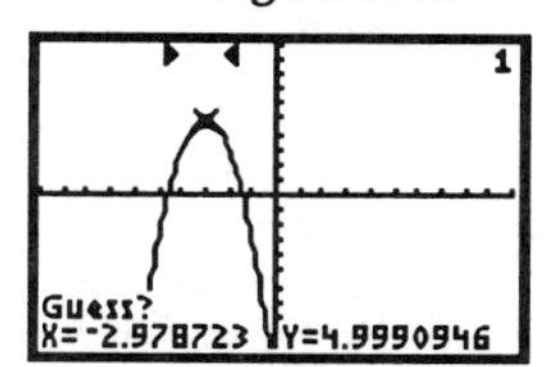

Figure 8. 19

8.1.5 The Vertex Form of a Quadratic Function.

We write the vertex as the point (h, k). Then the quadratic equation $y = ax^2 + x + c$ is transformed to :

$$y = a(x - h)^2 + k$$

So $y = -2x^2 - 12x - 13$ becomes

$$y = -2(x - {}^-3)^2 + 5$$
$$= -2(x + 3)^2 + 5$$

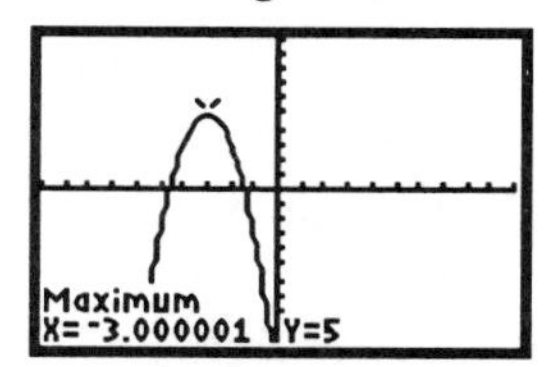

Figure 8. 20

Verify algebraically:

$$y = -2(x + 3)^2 + 5$$
$$= -2(x^2 + 6x + 9) + 5$$
$$= -2x^2 - 12x - 18 + 5$$
$$= -2x^2 - 12x - 13$$

Verify Numerically:

Enter both equations into [Y=] . See Figure 8.21. Set up a table of values. Press [2nd] [TblSet] . See Figure 8.22. Press [2nd] [TABLE] . Figure 8.23 shows that for all values of x , Y1 and Y2 are equivalent.

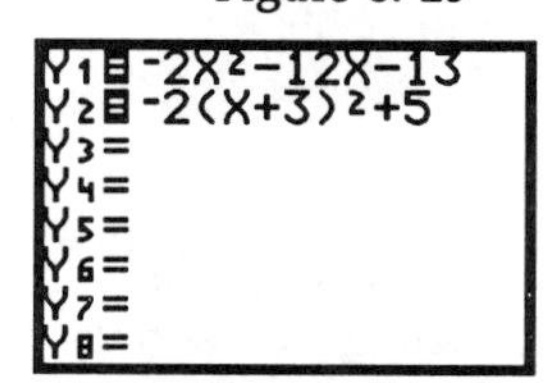

Figure 8. 21

Figure 8. 22

Therefore Y1 = Y2, or

$$-2x^2 - 12x - 13 = -2(x + 3)^2 + 5 \text{ and}$$
$$ax^2 + bx + c = a(x - h)^2 + k$$

X	Y1	Y2
-5	-3	-3
-4	3	3
-3	5	5
-2	3	3
-1	-3	-3
0	-13	-13
1	-27	-27

X= -3

Figure 8. 23

8.1.6 Finding an Appropriate Window.

Example 3

Graph $y = 3x^2 - 20x + 45.$

Into Y1 enter: $3x^2 - 20x + 45.$

CLEAR all other functions.

Press ZOOM 6 . We see nothing!

X	Y1	
-5	220	
-4	173	
-3	132	
-2	97	
-1	68	
0	45	
1	28	

X= -5

Figure 8. 24

Use the table to get a feel for what happens to y as x increases. Press 2nd TABLE . Figure 8.24 shows that when $x = -5$, $y = 220$ and when $x=0$, $y= 45$. No wonder we couldn't see anything on a [-10,10] by [-10,10] standard window.

From algebra we know that $a = 3 > 0$ so the graph opens upward. From the table use down arrow to find the minimum value of y (the vertex). See Figure 8.25, it looks like the minimum occurs around $x = 3$.

X	Y1	
0	45	
1	28	
2	17	
3	12	
4	13	
5	20	
6	33	

X=3

Figure 8. 25

Adjust the window so that you can see the point (3, 12) as well as the point (-5, 220). Press WINDOW ∇ ; let Ymax = 300 . See Figure 8.26.

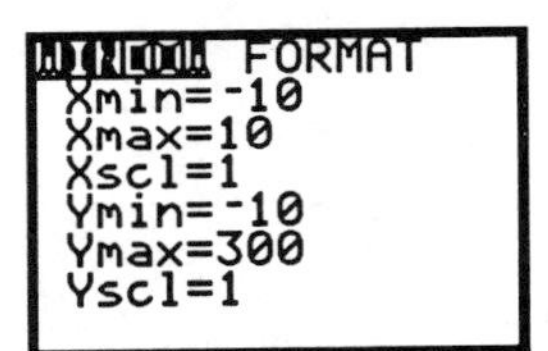

Figure 8. 26

Press GRAPH . See Figure 8.27. It looks like the graph is being cut off on the right side. We need to see more values of x. Press WINDOW . Change Xmax to 15.

Press GRAPH . See Figure 8.28.

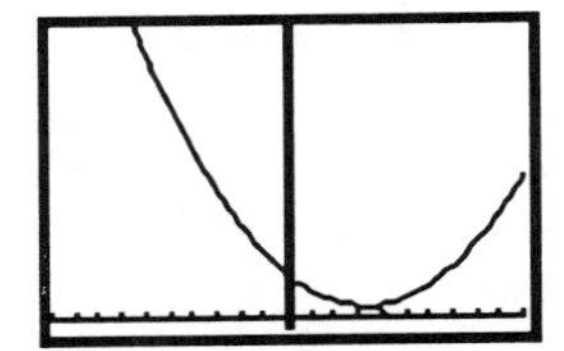

Figure 8. 27

8.1.7 A Complete Graph

Try to select a window that displays a *complete graph*. A complete graph shows the whole shape of the graph, with all its turning points and end behavior. Also shown are the y -intercept and x -intercept(s), if they exist. There are many complete graphs.

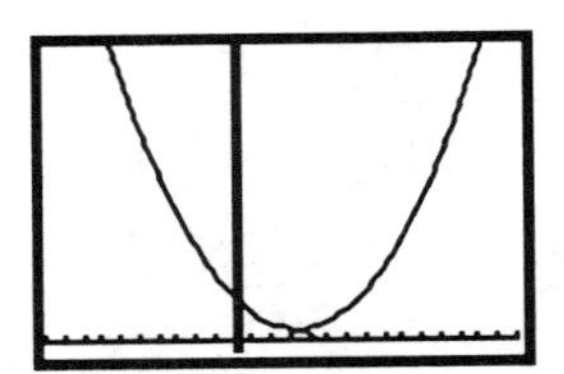

Figure 8. 28

Figure 8.28 shows a complete graph of

$$y = 3x^2 - 20x + 45$$

8.2 Polynomial Functions

In Chapter 9 you were introduced to power functions. When positive integer power functions are added together you get a polynomial function:

$y = a_n x^n + a_{n-1} x^{n-1} + \ldots + a_1 x^a + a_0$

where a_n is a constant coefficient ($a_n \neq 0$), and n is a positive integer power.

L1	L2	L3
1986	8663	------
1987	8986	
1988	9416	
1989	9780	
1990	9936	
1991	10050	
1992	9977	

L1(1)=1986

Figure 8. 29

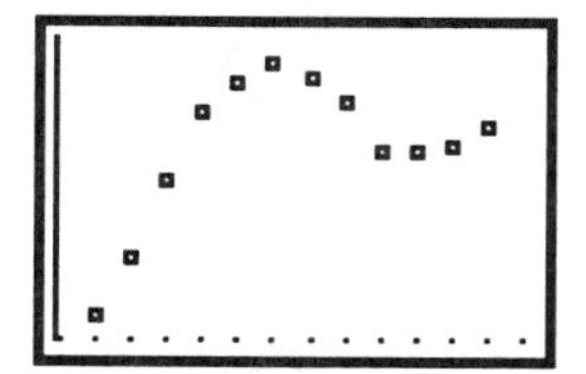

Figure 8. 30

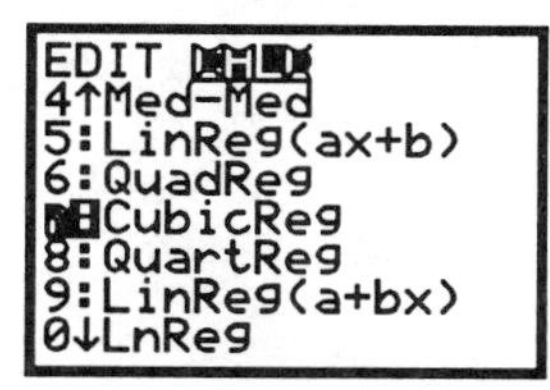

Figure 8. 31

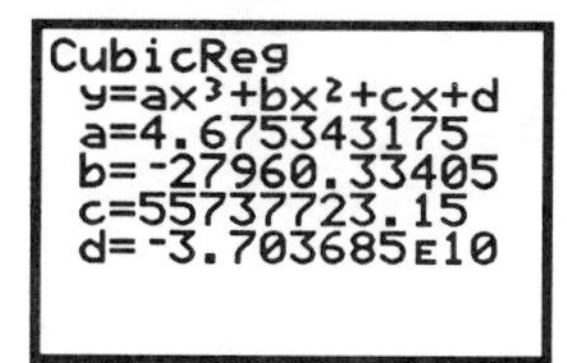

Figure 8. 32

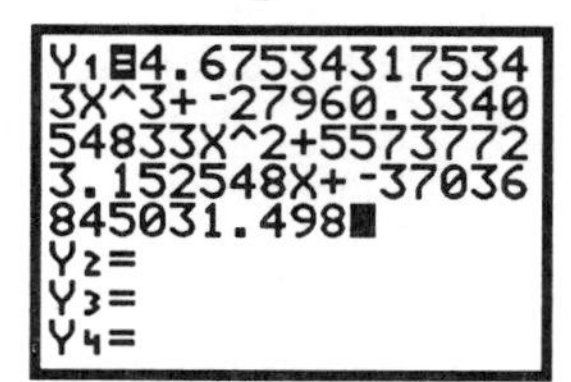

Figure 8. 33

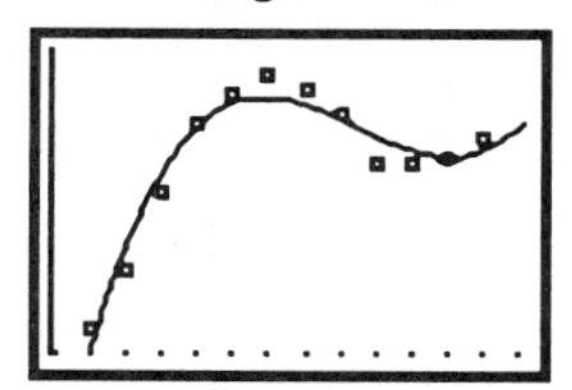

Figure 8. 34

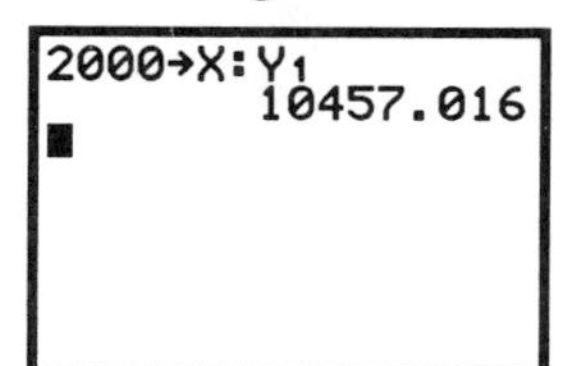

Figure 8. 35

Example 4

The fall term enrollment of a university's freshman across the years is given below. Find a regression model that would predict the enrollment in 2000 if trends continue

Table 8. 1

year	freshman
1986	8662
1987	8986
1988	9416
1989	9780
1990	9936
1991	10050
1992	9977
1993	9846
1994	9570
1995	9582
1996	9610
1997	9693

Press [STAT] [1:Edit]. Enter the years into L1 and the number of freshman in L2 . See Figure 8.29. Press [ZOOM] [9:ZoomStat] . See Figure 8.30. The data appears to increase to 10,050 then decrease to 9,570 then increase again. A polynomial regression equation with at least two turning points is a cubic.

Find the cubic regression model. Press [STAT] <CALC>; select [:CubicReg]. See Figure 8.31. Press [2nd] [L1] [,] [2nd] [L2] [ENTER] . Figure 8.32 shows all the coefficients of the third degree polynomial, or cubic function. Store the equation to Y1. Press [Y=] [VARS] ; select[5:Statistics] <EQ> ; select [:RegEQ]. Press [GRAPH] . See Figs. 8.33 and 8.34. To find the predicted number of freshman in 2000, store 2000 to x to find the value of Y1=10457. See Figure 8.35.

Extended Exploration 2:
The Mathematics of Motion: Freefall Data

EE2.1 Free Fall Data

Can you think about how the height of an object varies over time when the object is dropped from a location like a window or bridge or even from your hand? If performing an experiment like dropping a ball in a classroom, the CBL™ (Calculator Based Laboratory) can be used. Instructions for using the CBL™ will be shared with you by your instructor or you can contact the WEB site: http://www.ti.com

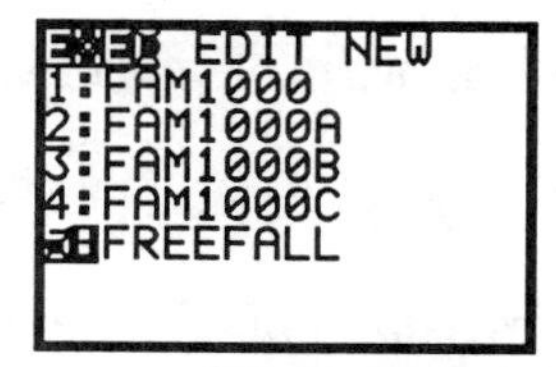

Figure EE2. 1

L1	L2	L3
0	0	87.05
.0167	1.72	85.33
.0333	3.75	83.3
.05	6.1	80.95
.0667	8.67	78.38
.0833	11.58	75.47
.1	14.71	72.34

L3(1)=87.05

Figure EE2. 2

Note: Be sure to review EE1.1.1 Receiving Data.

The text comes with a free fall data program that can be transferred from calculator to calculator. Once the data is in your calculator either by experiment or from the program, do the following.

1. Plot the data.
2. Determine a quadratic regression equation for the data.

Figures EE2.1 to EE2.6 represent the calculator steps you need to perform.

Figure EE2. 3

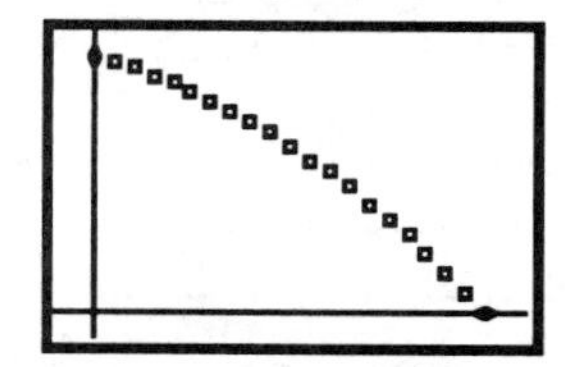

Figure EE2. 4

Note: Remember to execute the program before you set up your plot. Press PRGM ; select FREEFALL ENTER . The program stores data to L1 - L2. See Figure EE2.2.

You can also type the FREEFALL program (see below) into your calculator. Press PRGM <NEW> ENTER ; type the program name; ENTER ; and then type each line of the program. Refer to your manual for more specifics. (See also the subsequent table)

Figure EE2. 5

Figure EE2. 6

```
PROGRAM: FREEFALL
:ClrList L1,L2,L3,L4,L5,L6
:{0.0000,0.0167,0.0333,0.0500,0.0
667,0.0833,0.1000,0.1167,0.1333,0
.1500,0.1667,0.1833,0.2000,0.2167
,0.2333,0.2500,0.2667,0.2833,0.30
00,0.3167,0.3333,0.3500}→L1
:{0.00,1.72,3.75,6.10,8.67,11.58,
14.71,18.10,21.77,25.71,29.90,34.
45,39.22,44.22,49.58,55.15,60.99,
67.11,73.48,80.10,87.05,94.23}→L2
:{94.23,87.05,85.33,83.30,80.95,7
8.38,75.47,72.34,68.95,65.28,61.3
4,57.15,52.60,47.83,42.83,37.47,3
1.90,26.06,19.94,13.57,6.95,0.00}
→L3
```

This data is available as a GraphLink File for the TI-82 or TI-83, and as an Excel file, available at: http://www.wiley.com/college/Kimeclark.

Extended Exploration: The Mathematics of Motion

Distance fallen over time by an object in free fall

Time	Distance fallen	Distance from the Ground	
(sec)	(cm)		
0.0000	0.00	94.23	
0.0167	1.72	87.05	
0.0333	3.75	80.10	
0.0500	6.10	73.48	
0.0667	8.67	67.11	
0.0833	11.58	60.99	
0.1000	14.71	55.15	
0.1167	18.10	49.58	
0.1333	21.77	44.22	
0.1500	25.71	39.22	
0.1667	29.90	34.45	
0.1833	34.45	29.90	
0.2000	39.22	25.71	
0.2167	44.22	21.77	
0.2333	49.58	18.10	
0.2500	55.15	14.71	
0.2667	60.99	11.58	
0.2833	67.11	8.67	
0.3000	73.48	6.10	
0.3167	80.10	3.75	
0.3333	87.05	1.72	
0.3500	94.23	0.00	

Part II

Student Solutions Manual

Introductory Notes:

This solutions manual contains answers to the odd-numbered exercises found at the end of each chapter of the second edition of *Explorations in College Algebra*.. Deliberately, the answers given here are much more full than those found at the back of the text. They should help you understand better what is expected of you in your answers.

Some notation conventions that you need to know:

a. The symbol "•" is often used to indicate multiplication between numeric constants and/or algebraic symbols. Thus, *e.g.*, 10•50 evaluates to 500 and u•v is the same as uv.

b. Because of certain limitations in the software used to produce many of the graphs in this manual, instead of the usual exponential notation, the symbol "^" had to be used in writing formulae in these graphs. Thus, *e.g.*, 2^4 is used instead of 2^4 and evaluates, of course, to 16. A similar problem exists in trying to produce subscripts. Thus, *e.g.*, log3(25) had to be used formulae written in the graphs instead of the usual $\log_3(25)$

c. Occasionally, some additional commentary is judged to be needed in order to have you better understand the answer given. Each such addition is put in "[..]"'s and is labeled with "**COMMENT**".

d. The end of the solutions to each chapter is designated by a line of <><>'s.

Lastly, while these solutions have been proofread and checked several times by several persons, there could still be mistakes. I would appreciate being informed of any errors that have, until now, gone undetected. Any suggestions for improving the solutions are also welcome. In addition, each chapter's solution set is made to begin on an odd-numbered page so that you can tear out the solution pages for that chapter as a whole and either send it to me with your comments or use it as a separate booklet in your work.

J. A. L.

Part II: Student Solutions Manual

Contents

Exercises for Section 1.1

1. a. Mean $= 386/7 = 55.14$; median $= 46$.

b. Changing any entry in the list that is greater than the median to something still higher will not change the median of the list but will increase the mean. The same effect can be had if an entry less that the median is increased to a value that is still less than or equal to the median.

3. The mean annual salary is \$24,700 and the median annual salary is \$18,000. The mean is heavily weighted by the two high salaries. The mean salary is more attractive but is not likely to be an accurate indicator.

5. No answer is given here. (In general, when your answers can vary quite a bit, either a typical answer or none is given.)

7. He is correct provided the person leaving state A has an I.Q. that is below the average I.Q. of people in state A and above the average I.Q. of the people in state B.

9. The mean net worth of a group, *e.g.*, American families, is heavily biased upwards by the very high incomes of a relatively small subset of the group. The median net worth of such groups is not as biased. The two measures would be the same if net worths were distributed symmetrically about the mean.

11. a. $(\sum_{i=1}^{5} x_i)/5$ **b.** $(\sum_{i=1}^{n} t_i)/n$ **c.** $\sum_{k=1}^{5} 2k = 30$

13. Using the midpoint of each age grouping as the representative age of that class (*i.e.*, 4.5, 14.5, ... , 84.5), the resultant mean age is $9784297.5/272635 = 35.888$ yrs. If instead of the actual midpoints one takes the "natural values"(*i.e.*, 5, 15, ... , 85) to be the representative ages, then the mean age is $9920615/272635 = 36.388$ years.

15. a. 44

b. Quantitative; ranges of money in units of \$1000 are provided.

c. $10/44 = 22.7\%$

d. (The chart is given on the next page)

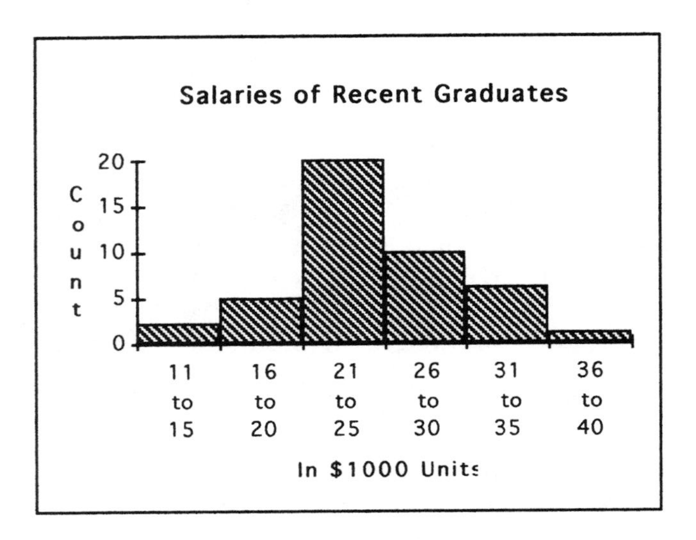

17. a. Housing ; 41.4%

b. $5220.00

c. A general coverage theme might be: "Where does the average American household's money go? The pie chart in the text gives an answer for late 1992." One could also have a particular theme such as: "Americans spend more on housing than on anything else!". Any accurate observation about some particular category would also be acceptable.

19. The mean is 162/50 = $3.24 and the median is $3

Exercises for Section 1.2

21. Your answers will probably be different. One very noticeable trend is the increase in 2050 of the percentages of persons in each 5-year age spread, starting at 45 and going up. Another trend is the vanishing of dominance by the 20-to-40 year old group. The projections for 2050 show a fairly even distribution among age groups from infancy up to 40.

23. a. Northeast: mean = 204.9/9 = 22.77; median = 9.20
Midwest: mean = 238.9/12 = 19.91; median = 18.05

b. Your summary paragraphs could mention the closeness of mean and median for occupancy rates in NE and MW and the great discrepancy between these for personnel in both sections as well as a great discrepancy between these for beds in the NE. They could also mention the widely differing numbers of beds and personnel in various states in the NE and NW *vs.* the relative homogeneity of occupancy rates in both regions.

25. Comments could include statements like the following: The population of China is substantially younger than that of the U.S. In 1990 the median age in China was 25 and the mean was 28. In contrast, the median age in the U.S. was 33 and the mean age was 36. As the graphs indicate, the largest percentage of China's population is in the interval from 20 to 24, representing approximately 11.1% of the population. In the US, the highest percentage, 8.8%, was for people from 30 to 34. Below age 30, China had a larger percentage than the U.S. in every 5-year interval.

27. a. Sex is a qualitative variable; age and weight are quantitative variables.

b. Below is a table with the computed data -- all entries are measured in pounds.

1993 data	Weights of Wolves in the Northwest	
	Mean	Median
Male	103.3	98
Female	68.5	72

c. The median is probably the best measure of central tendency for the males because of the presence of an outlier. The females do not seem to have an outlier and thus either measure will do.

d. There are 4 male wolves and 15 females but only 14 have age designations. If we count a pup's ages as 0.5 years and take the midpoint when a range is given, then the mean age of the 4 males is 3.63 years; the mean age for the 14 females with ages given is 1.96 years; the mean age overall is 2.33 years.

e. Your summaries will probably be different.

29. a. i. The number of males in the US in 1997 in the 35 - 39 age range is approximately 11.4 million.

ii. The number of females in the US in 1997 in the 55 - 59 age range is approximately 6.1 million

iii. The number of males, in the US, in 2050, in the 85+ age category, is projected to be approximately 7.1 million. The number of females in the US, in that age category, in 2050, is projected to be approximately 12 million.

iv. The total US population in 2050 in 0-9 age range is projected to be approximately 53.4 million.

[**COMMENT**: Estimates were made using a centimeter ruler. Your eyeball estimates will probably be different.]

b. Typical observations: in 2050 there will be a huge increase in the 85+ age range. The largest group in 1997 was in the 35-39 age range. In 2050 all of the 5 year age ranges below 50 will be larger.

31. This is open to interpretation. By 2050 the mean age may be greater than the median for the United States, as it is more affected than the median by outlier values (*e.g.*, by those living longer because of medical advances). On the other hand, the pyramid seems as if it will be weighted downward and this could bring the mean value down. These guesses may or may not be true in smaller communities in the US. It may also not be true in certain impoverished areas where medical care is not that good. You will probably say much the same.

Exercises for Section 1.3

33. a. Increased in 86 -87 and 89 - 93.
b. Decreased during 87-89 and 93-94
c. $12 billion maximum in sales in 1993
d. $7.5 billion minimum in sales in 1989

35. a. His maximum temperature during that day was approximately 37.4 °C.
b. His minimum temperature during that day was approximately 36.3 °C.
c. His temperature was fairly the same from noon to 6 PM and then started going down until it reached a low around 1 AM, then slowly rose until 7 AM and finally crept up to where it was around noon on the previous day. (Your focal points will probably be different from these.)

37. a. Johnsonville's population goes from 2.4 x 100,000 to 5.8 x 100,000. Palm City's population ranges from 1.8 x 100,000 to a high of 3.8 x 100,000. (Note that this notation adheres to what is found in the graph.)
b. The population of Palm City increased from 1900 to 1930
c. The population of Palm City decreased from 1930 to 1990.
d. The two populations were equal sometime around 1938.

Exercises for Section 1.4

39. a. Add 1 to the value of x; divide the result by what one gets by subtracting 1 from the value of x.
b. (5, 1.5) **c.** (2,3) **d.** No, the formula is not defined if $x = 1$.

41. a. $x = 0$ implies $y = 0$ **b.** If $x > 0$ then $y < 0$. **c.** If $x < 0$ then $y < 0$. **d.** No

43. a. Only (-1,3) solves $y = 2x+5$. **b.** (1,0) and (2,3) solve $y = x^2 - 1$.
c. (-1,3) and (2,3) solve $y = x^2 - x + 1$ **d.** Only (1,2) solves $y = 4/(x+1)$.

Exercises for Section 1.5

45. a. $S_1 = 0.90 \bullet P$; $90 **b.** $S_2 = (0.90)^2 \bullet P$; $81

c. $S_3 = (0.90)^3 \bullet P$; $72.90 **d.** $S_5 = (0.90)^5 \bullet P$; $59.05; 40.95%

47. Yes, the graph represents a function of x. It passes the vertical line test.

49. No, height is not a function of weight because an input of 120 results in two different outputs. Weight, however, is a function of height because there are no repeated entries in the weight column.

51. a. The graph labeled (i) describes a function.
b. It passes the vertical line test; the other two do not.

53. a. True; each weight input has only one output cost.

b. False; *e.g.*, the cost to mail a 3 oz. parcel can be \$3.59 or \$3.90, depending on the zone. Weight, however, is a function of cost if one sticks to a single zone.

c. True; the cost for all zones is the same.

d. True; the same weight can be mapped to different zones.

55. a. The formulae are: $y = x + 5$; $y = x^2 + 1$; $y = 3$

b. All 3 represent y as a function of x; each input of x has only one output y.

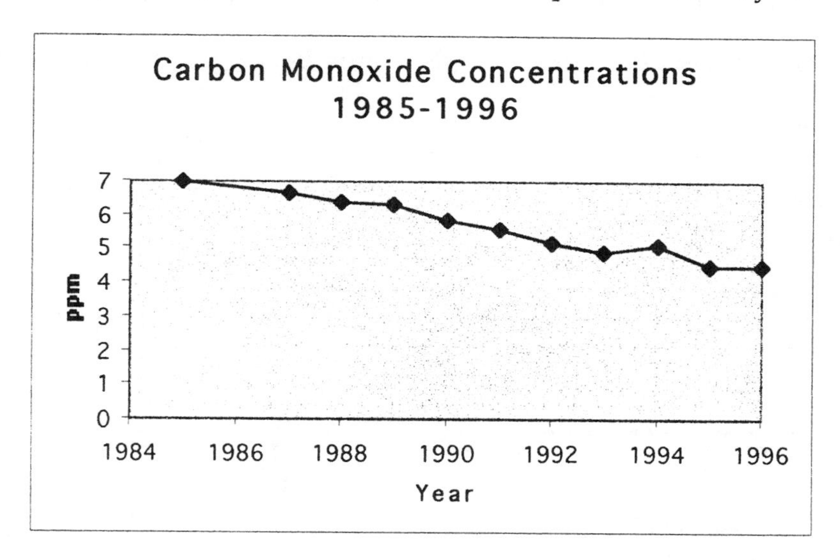

57. a. The set of years would make the logical choice of independent variable and the ppm of carbon monoxide is the logical dependent variable. The chart given above shows what happens to the ppm of carbon monoxide over the years.
b. The carbon monoxide ppm is a function of the year since each year is associated with only one level of carbon monoxide.
c. The carbon monoxide ppm content varies with the year.

d. The graph is given on the previous page. The data points were plotted and they were connected with straight line segments in order to communicate better the changes over time.

e. The function is decreasing from 1985 to 1993 and from 1994 to 1996; it is increasing from 1993 to 1994

f. 6.97 ppm is the maximum value and 4.20 ppm is the minimum value.

59. a. Yes, P is a function of Y since the inputs are all distinct.

b. The domain is the set of years from 1990 to 1995 inclusive; the range is the set of corresponding money values, namely { -0.5, 0, 1.2, 1.4, 2.3 }

c. The maximum P value is 2.3; it occurs when Y = 1991

d. P is increasing from 1990 to 1991 and from 1993 to 1994

e. Y is not a function of P, since the two inputs of 1.4 have different outputs.

61. a. $C = 1.24 \bullet G$, where C is cost in dollars and G = number of gallons purchased.

b. G is the independent variable and C the dependent one.

c. For each input of G, there is one and only one value for C; thus it is a function.

d. A suitable domain would be from 0 to the maximum capacity in gallons of a gas tank. We can play it safe and say 50 gallons. The corresponding range is then from \$0 to \$62. Clearly if there are larger tanks on the market, the domain and range will adjust accordingly.

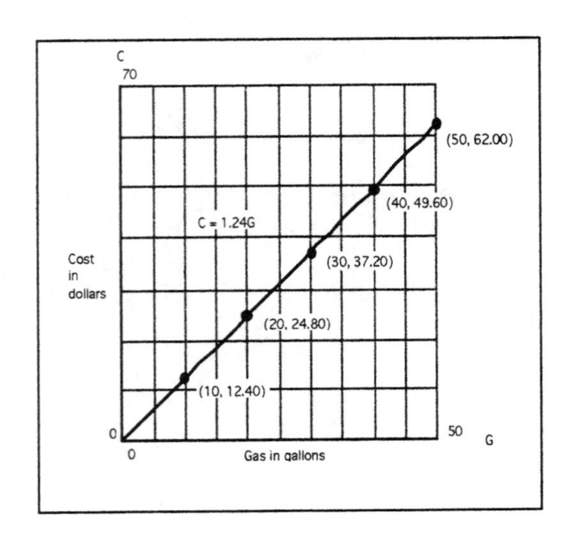

e. Here is a small table of values:

Capacity	Cost
10	12.40
20	24.80
30	37.20
40	49.60
50	62.00

This table and the graph of the function are illustrated in the diagram given on the left.

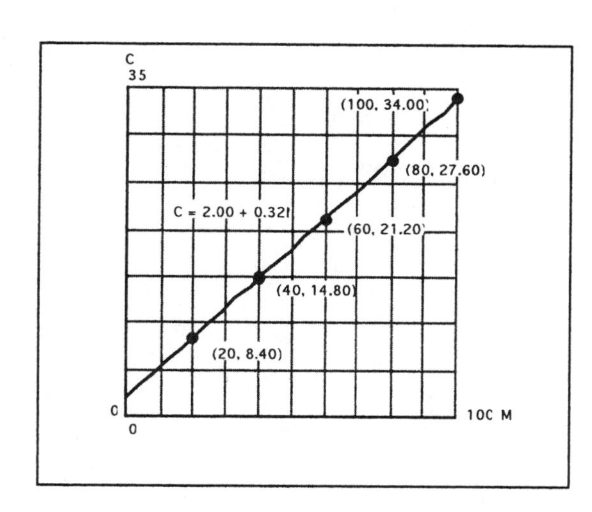

63. The equation is C= 2.00 + 0.32M. It represents a function. The independent variable is M. The dependent variable is C. Its graph is to the left. Here is a table of values:

Miles	Cost	Miles	Cost
0	2.00	30	11.60
10	5.20	40	14.80
20	8.40	50	18.00

Some of these values are marked in the graph on the left.

65. a. f(2) = 4 **b** f(-1) = 4 **c.** f(0) = 2 **d.** f(-5) = 32.

67. a. p(-4) = 0.063, p(5) = 32 and p(1) = 2. **b.** n = 1 only

69. a. f(-2) = 5, f(-1) = 0, f(0) = -3 and f(1) = -4.

b. f(x) = -3 if and only if x = 0 or 2.

c. The range of f is from -4 to ∞ since we may assume that its arms extend out indefinitely.

71. f(0) = 1, f(1) = 1 and f(-2) = 25

73. a. 2 **b.** -8 **c**. 10 **d**. -5

<><><><><><><><><><><><><><><><><><><><><><><><><><><><><><><><><><><><><><><><>

Exercises for Section 2.1

1. a. (423 - 466)/(1994 - 1967) = -43/27 = -1.59 points per year
b. (423 - 423)/(1994 - 1992) = 0/2 = 0.00 points per year

3. a. (8,049,875 - 631,983)/ (1998 - 1985) = 7417892/13 = 570,607.07 computers/year
b. (6.4 - 62.7)/(1998 - 1985) = -56.3/13 = -4.33 students per computer per year

5. a. (83.7-26.1)/(1998-1940) = 57.6/58 or 0.99 is the average rate of change per year from 1940 to 1998 for whites . For blacks the average rate is (76.0 - 7.3)/58 = 68.7/58 ≈ 1.18. For Asian Pacific Islanders this average is (85.0 - 22.6)/58 = 62.4/58 ≈ 1.08. For all this average is (82.8 - 24.5) / 58 = 58.3/58 ≈ 1.01. Each of these percents measures the average change in percent per year over the 58 years.

b. If these rates continue: the percentage in 2000 for whites will be 83.7 + 2•0.99 = 85.68, for blacks it will be 76.0 + 2•1.18 = 78.36 and for Asian/Pacific Islanders it will be 85.0 + 2•1.08 = 87.16 For all, it will be 82.8 + 2•1.01 = 84.82.

c. Comments are expected about the great increase in the percent of each group of persons over the past 58 years. You might also mention that blacks have had the greatest average rate of increase over this period with 1.18% per year.

d. For whites: solving 83.7 + x•0.99 = 100 for the number of years x since 1998 gives x ≈ 16.46 or in mid 2015. For blacks: 76.0 + x•1.18 = 100 gives x ≈ 20.34 or early in 2019. For Asian/Pacific Islanders: 85.0 + x•1.08 = 100 give x ≈ 13.89 or near the end of 2012. For all, 82.8 + x•1.01 = 100 gives x ≈ 17.03 or early in 2016.

7. a. In 1900, white females had the highest average life expectancy. It is expected that white females will have the highest life expectancy in 2000. In 1900, black males had the lowest average life expectancy. The same is expected to be true for them in 2000.

b. The average rate of change in life expectancy from 1900 to 2000 for each of the four groups is as follows:

white males: (74.2 - 46.6)/100 = 0.276; white females: (80.5 - 48.7)100 = 0.318;
black males: (64.6 - 32.5)/100 = 0.321; black females: (74.7 - 33.5)/100 = 0.412.

Thus black females had the highest average rate of change in life expectancy from 1900 to 2000.

c. Your summary could easily include the observations made in **a.** and **b.** In general, all groups mentioned have enjoyed great increases in life expectancy during the past century. In only one case since 1950 has a group dropped in life expectancy from one decade to the next. Black men fell slightly from 60.7 years to 60.0 years between 1960 and 1970. Possible reasons could include urban poverty and overrepresentation among the infantry in Vietnam.

Exercises for Section 2.2

9. a.

Year	Salary ($ millions)	Rate of Change over prior year
1990	0.60	n.a.
1991	0.85	0.25
1992	1.03	0.18
1993	1.08	0.05
1994	1.17	0.09
1995	1.11	-0.06
1996	1.12	0.01

b. The average rate was smallest in the 1995-1996 period. The slope of the line segment joining the two data points is the least steep.

c. The average rate was greatest in the 1990-1991 period. The slope of the line segment joining these two points is the steepest.

d. This summary could give the range in the average salaries from 1990 to 1996; the graph shows that things have leveled off a bit since 1994; you might point to how large these average salaries are even if the annual increase has slowed down quite a bit.

[COMMENT: Steepness is measured here in terms of absolute values of the slopes computed. Thus the smallest average rate of change is not -0.06 but 0.01 even though -0.06 is smaller numerically than 0.01. As stated, the line is less steep if the slope is 0.01 than if it is -0.06.]

11. a. From 1850 to 1950. The average rate of change for this period is $(30.2\text{-}18.9)/100 \approx 0.113$ years of age per year.

b. $(22.9 - 18.9)/(1900\text{-}1850) = 4.0/50 = 0.080$; $(30.2 - 22.9)(1950 - 1900) = 7.3/50 = 0.146$; $(35.7 - 30.2)(2000 - 1950) = 5.5/50 = 0.110$ and a projected: $(38.1 - 35.7)/(2050 - 2000) = 0.048$ -- all measured in years per calendar year.

c. Possible causes are: better health care, better delivery of food and better nutrition or a shrinking birth rate, caused, in part, by the rise in the use of birth control. Since the median is the statistical middle value, it is rather likely that the increase in median age implies that people are living longer but there is no necessity that this is the case.

d. The table asked for is given on the next page. The units in the 3rd column are years per calendar year.

e. The plot of the first and third columns is given on the next page.

f. Negative rates of change show up in the 1950-to-1960 and 1960-to-1970 periods. During those two periods there were two major wars involving the US. that took the lives of many men. Other causes might be increase in birth rate and immigration of young people.

Table for **#11. d.**

Med. Age of	US Population	1860-1990
Year	Median Age	Average Rates of Change
1850	18.9	n.a..
1860	19.4	0.050
1870	20.2	0.080
1880	20.9	0.070
1890	22.0	0.110
1900	22.9	0.090
1910	24.1	0.120
1920	25.3	0.120
1930	26.4	0.110
1940	29.0	0.260
1950	30.2	0.120
1960	29.5	-0.070
1970	28.0	-0.150
1980	30.0	0.200
1990	32.8	0.280
2000	35.7	0.290

Graph for **#11. e.**

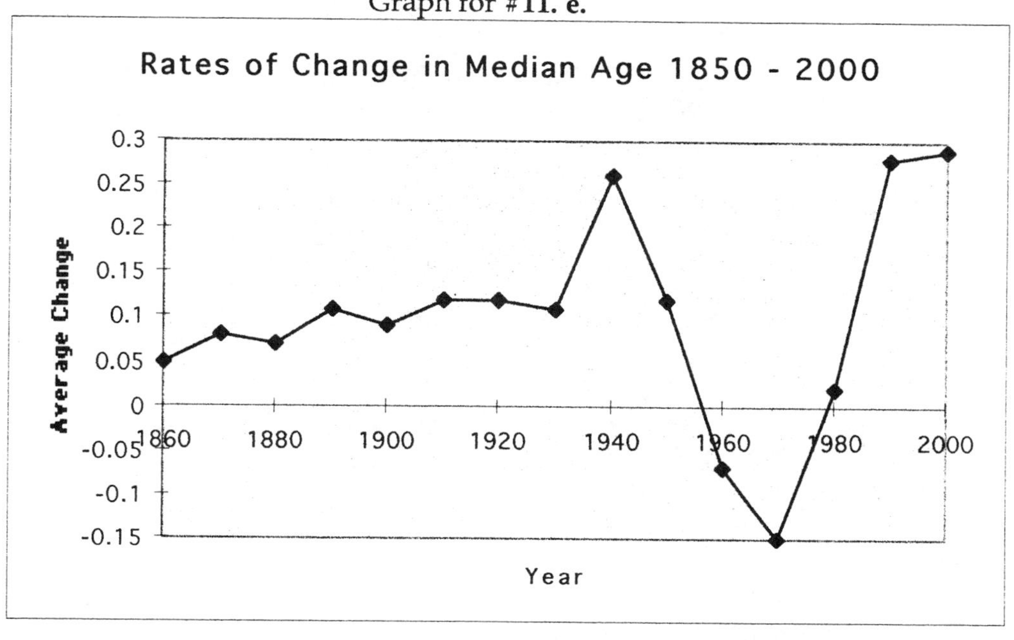

13. a In 1920 there were 27,791 million papers printed and there were approximately 105.7 million persons. Thus on average there were 0.26 papers published per person. Here is the math: $(27791 \bullet 10^3)/(105.7 \bullet 10^6) \approx 0.26$. In 1990 there were 63,324 million newspapers printed and there were approximately 248.7 million people. Hence there were $(63324 \bullet 10^3)/(248.7 \bullet 10^6) \approx 0.25$ papers per person. Thus, while there are a lot more newspapers printed, the distribution is roughly the same: approx. 1 newspaper is printed for each 4 persons in this country.

b. The table below indicates the average rates of change for each 5-year period since 1950 for newspapers published and for TV stations for each decade. The graphs below illustrate this table.

Year	Avg. Annual Change in TV Stations	Avg. Annual Change in Newsp. Publ
1950	n.a.	n.a.
1955	62.6	-2.4
1960	20.8	0.6
1965	10.8	-2.4
1970	21.6	-0.6
1975	5.8	1.6
1980	5.6	-2.2
1985	29.8	-13.8
1990	41.8	-13
1995	88.0	-5.6

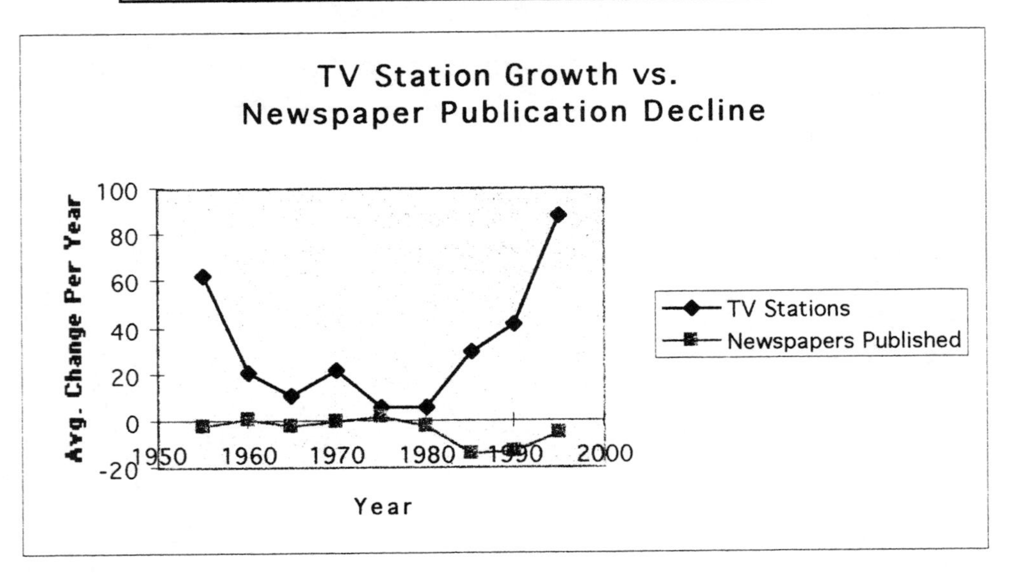

c. There would be $1532 + 88 \bullet 10 = 2412$ TV stations in 2005, if the growth rate from 1990 to 1995 prevailed. It seems too large

d. You might make comments like the following. Fewer newspapers are surviving (while the readership of these fewer goes up). Meanwhile, the number of on air TV stations has soared. Nonetheless, there is a glaring absence of two types of data here and thus talking of trends is difficult. First, we do not know what sorts of channels dominate this recent proliferation. Perhaps these are mostly movie channels and pay-per-view stations with no news. Second, and more importantly, while we have statistics on newspaper copies printed and thus some information on readership, we have no corresponding data about TV news viewership.

Exercises for Section 2.3

15. **a.** $-4 = (1 - t)/(-2 - 3)$ or $20 = 1 - t$ or $t = -19$
b. $2/3 = (9 - 6)/(t - 5)$ or $2(t - 5) = 3(9 - 6)$ or $2t - 10 = 9$ or $t = 9.5$

17. The point sequences are collinear if the slopes of the line segments between each successive pair have the same value.

a. The slope of the line segment joining (2,3) and (4,7) is $(7 - 3)/(4 - 2) = 4/2 = 2$
The slope of the line segment between (4,7) and (8,15) is $(15 - 7)/(8 - 4) = 8/4 = 2$
Thus these three points ARE collinear.

b. The slope of the line segment between (-3,1) and (2,4) is $(4 - 1)/(2 + 3) = 3/5$
The slope of the line segment between (2, 4) and (7,8) is $(8 - 4)/(7 - 2) = 4/5$.
Thus these three points are NOT collinear.

19. **a.** Positive over $B < x < F$; negative over $F < x \leq H$; zero over $A \leq x \leq B$ and at $x = F$.
b. Positive over $B < x < C$ and $D < x < E$; negative over $E < x < G$; zero over $A \leq x \leq B$, $C < x < D$ and $G < x \leq H$. (Some dividing points are corners and at these the slope is not uniquely defined.)

21. Answer is omitted (see the introduction).

Exercises for Section 2.4

23. **a.** One encouraging fact is that the number of cases reported went down from 1991 to 1992 by 385 cases and again from 1993 to 1994 by 2341 cases -- the drop being more than 6 times as great.

b. The most discouraging fact is the sharp rise in the 1992-1993 period -- more than 15 times as great as the rise from 1990 to 1992. Also, the drop mentioned in part **a.**, while significant, still leaves the number of cases each year well above what it was for the 1986 to 1992 period. Also, while there is a drop in the number of cases from 1994 to 1997 the rate of the drop is slower than from 1993 to 1994

c. What appears to be a sudden increase in the spread of AIDS could be due to increased spending in a new, free and well-advertised testing program, resulting in an increased number of diagnoses of existing cases.

25. The horizontal axis of time is the same in each of the four graphs. The differences in appearance are due to the changing of the vertical scales. The display range in **(a)** is from 0% to 70% and since the actual percentages plotted on the graph range from 54% to 64%, the graph gives the impression that black income is on the high side. The display range in **(b)** by contrast is from 50% to 120% and thus the actual range of percentages appears to be very low since 54 - 64% is low in this display range. The range in **(c)** is from 40% to 65% but it is made to take up twice the height of either of the first two. Thus the visual distance between the low of 54% and the high of 64% is much bigger in appearance than its relative size demands. This leads to small numerical changes being visually exaggerated. The opposite happens in **(d)**. There the vertical scale is shrunk to half the visual size of that in **(a)** and **(b)** and a quarter of the visual size of that in **(c)**. Hence the impression is given that there has been virtually no change in black income as a percentage of white income.

The scales used take advantage of the power of an image to convey sizes. That power is much stronger than the numbers themselves (which do not vary) in causing impressions to be formed for most persons.

27. Omitted

Exercises for Section 2.5

29. a. E(0) = 5000, E(1) =6000, and E(20) = 25000

b. (0, 5000), (1, 6000) and (20, 25000)

31. a. (5000, 0) is not a solution to either equation.

b. (15, 24000) is a solution to the second equation but not the first.

c. (35, 40000) is a solution to the first equation and the second.

33. $dollars = dollars + \frac{dollars}{year} \bullet (number\ of\ years)$

35. a. hours **b.** miles/gallon **c.** calories/gram of fat

Exercises for Section 2.6

37. a. Slope = 0.4, vert. intercept = -20

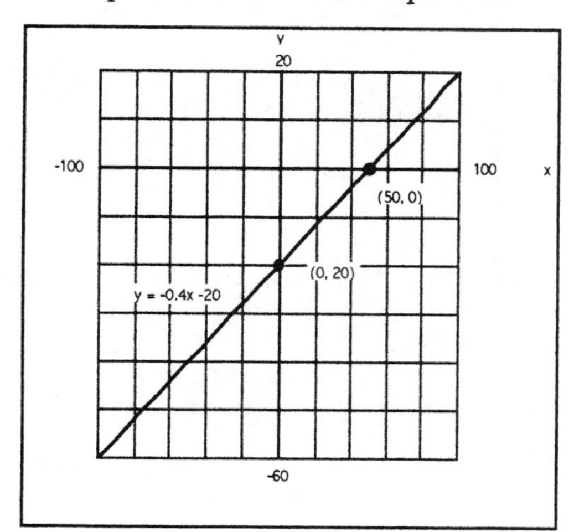

b. Slope = -200, vert. intercept = 4000

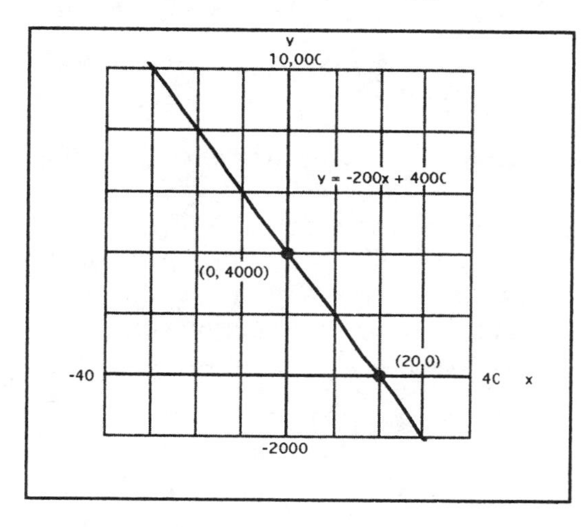

39. The equation is y = 3x –2. Its graph is given below on the left. A table is given on the right

x	y
-5	-17
-4	-14
0	-2

Graph for # 39

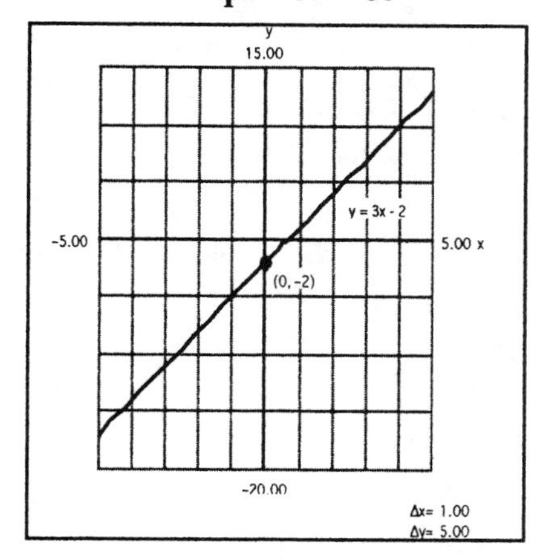

Graph for # 41

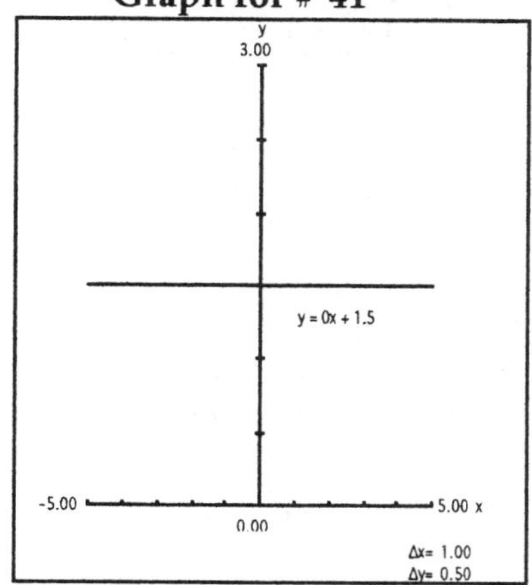

41. The equation is y = 0x + 1.5 The graph is given above on the right.

Small table:

x	y
-10	1.5
0	1.5
1	1.5

43. a. $y = 5x + 13$ **b.** $y = -0.75x + 1.5$ **c.** $y = 3$

45. a. $m = (5.1 - 7.6) / (4 - 2) = -2.5/2 = -1.25$ and $y - 7.6 = -1.25(x - 2)$ or $y = -1.25x + 10.1$
b. $m = (16 - 12) / (7 - 5) = 4/2 = 2$ and $W - 12 = 2(A - 5)$ or $W = 2A + 2$

47. Cost = 150 + 120•credits where credits = total number of credits registered for in all classes and 150 is measured in dollars and 120 in dollars per credit.

49. $C(n) = 2.50 + 0.10n$ where n = number of checks cashed that month.

51. a. Annual increase = (32000 - 26000)/4 = $1500
b. $S(n) = 26000 + 1500n$, where S(n) is measured in dollars and n in years from start of employment.
c. Here $0 \leq n \leq 20$ since the contract is for 20 years.

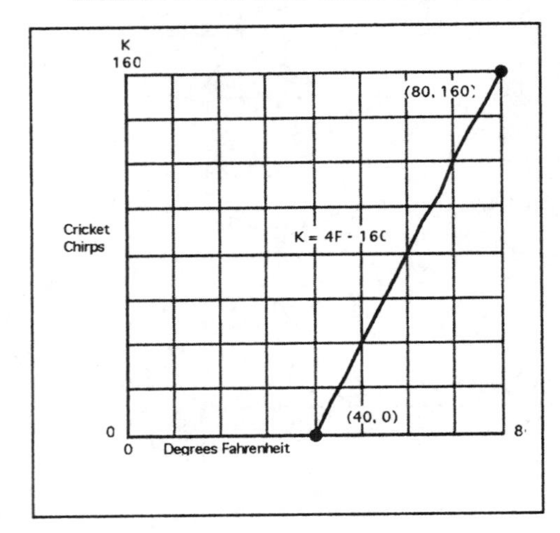

53. a. Slope = 4 and vertical intercept = -160.

b. Units for K and -160 are number of chirps per minute; units for F are degrees Fahrenheit; units for 4 are chirps per minute per degree Fahrenheit.

c. A reasonable domain of this model is from F = 40 to, say, F = 80. The lower value reduces the number of chirps to 0; the higher is just a guess. There is not enough information about the cricket's environment, etc., to choose better.]

d. A small table is given:

K	F
40	0
60	80
80	160

e. (160 -0)/(80 - 40) = 4; yes, it is what was expected because for a linear equation the rate of change between any two points is the same.
The graph is above on the left.

55. a. $P = 285 - 15t$

b. It will take 16.33 years to get the water safe for swimming since $40 = 285 - 15t$ implies that $15t = 245$ or $t = 16.33$ years.

57. a. A linear formula can fit the job if we let its slope be the average increase in the house's value per year. Here m = (250000 - 20000)/(1999 - 1973) = 230000/26 = 8846.15 dollars per year. The equation then is $V(t) = 20000 + 8846.15t$, where t measures years since 1973.

b. If the year is 2010, then $t = 37$ and $V(37) = \$347,307.55$.

c. $500000 = 8846.15t + 20000$ or $8846.15t = 480000$ or $t = 54.26$ years since 1973. If now means 1999 then in 1973 she was 57 - 26 = 31. In 54.26 years from 1973 she will be 85.26 years old.

59. The equation of the line illustrated is $y = (8/3)x - 4$. If one made the squares on the horizontal scale to be, say, 1/2 apart instead of 1 apart and kept the vertical scale the same as it is in the graph, then the graph would appear less steep since it would take twice as long (to the eye) to rise three units. If, instead, the vertical scale was made to be, say, 1/2 between lines instead of 1 and one kept the horizontal scale as it is in the text, then the graph would appear much more steep for it would seem to rise to twice the current height in the same x distance.

[**COMMENT**: The graphs are omitted because the graphing software employed does not allow the user a choice of distances between grid lines.]

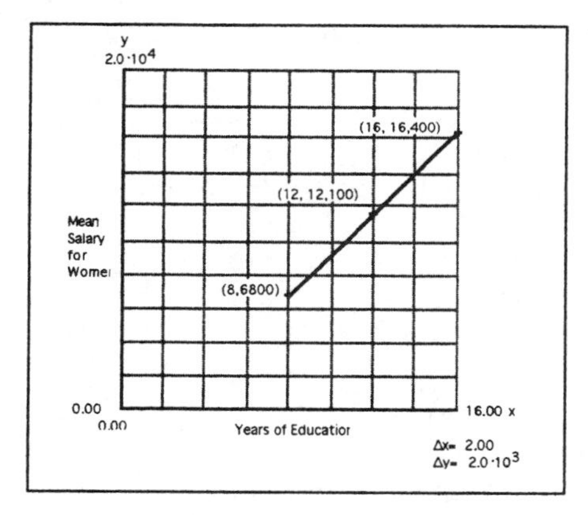

61. a. The data is plotted on the graph to the left. (A straight line was found to go through the three points and it was drawn.)

b. The plot suggests that the two variables are linearly related. This is confirmed by checking the slopes between the 1st and 2nd points and the 2nd and 3rd points:

$m = (11600\text{-}6800)/(12\text{-}8{=}4800/4{=}1200$; and
$m = (16400\text{-}11600)/(16\text{-}8) = 4800/4 = 1200$

c. $S - 6800 = 1200(E - 8)$ or $S = 1200E - 2800$, if $8 \leq E \leq 16$

63. The entries in the table argue that the relationship is linear. The average rate of change in salinity per degree Celsius is a constant: -0.054. Since FP for 0 salinity is 0 degrees. Celsius, we have that $FP = -0.054S$, where S is the salinity measured in ppt and FP, the freezing point, is measured in degrees Celsius.

Exercises for Section 2.7

65. a. The independent variable is the price P; the dependent variable is the sales tax T. The equation is $T = 0.065P$.

b. Independent variable is amount of sunlight S received; dependent variable is the height of the tree H. The equation is $H = kS$, where k is a constant.

c. Time t in years since 1985 is the independent variable and salary S in dollars is the dependent variable. The equation is $S = 25000 + 1300t$.

67. $d = 5t$; yes, d is directly proportional to t; it is more likely to be the person jogging since the rate is too slow for a car.

69. $D(t) = 15000$, where t is measured in years from 1983 and D(t) is measured in the number of flexible disk drives.

71. a. horizontal: $y = -4$; vertical: $x = 1$; slope 2: $y + 4 = 2(x - 1)$ or $y = 2x - 6$

b. horizontal: $y = 0$; vertical: $x = 2$ slope 2: $y - 0 = 2(x - 2)$ or $y = 2x - 4$

c. horizontal: $y = 50$; vertical: $x = 8$ slope 2: $y - 50 = 2(x - 8)$ or $y = 2x + 34$

73. a. Assuming a steady rate, it must be 10 lbs. per month.

b. $w(t) = 175$ where t = months until fall training and 175 is measured in lbs. The graph is a horizontal line. Of course, this is not the only way that his weight could be plotted as a function of time. It could have oscillated around 175; it could have grown and then went down. There is not enough information to say anything further.

c. If the independent variable were the days from end of Spring training instead of months, then the data would probably oscillate less but more frequently.

75. a. $y - 7 = -1(x - 3)$ or $y = -x + 10$
b. $y - 7 = 1(x - 3)$ or $y = x + 4$

77. For graph A: both slopes are positive; same y-intercept.
For graph B: one slope is positive, one negative; same y-intercept.
For graph C: the lines are parallel; different y-intercepts..
For graph D: one slope is positive, one negative; different y-intercepts.

79. a. $y = (-A/B)x + (C/B)$, $B \neq 0$

b. The slope is $-A/B$, $B \neq 0$

c. The slope is $-A/B$, $B \neq 0$

d. The slope is B/A, $A \neq 0$

Exercises for Section 2.8

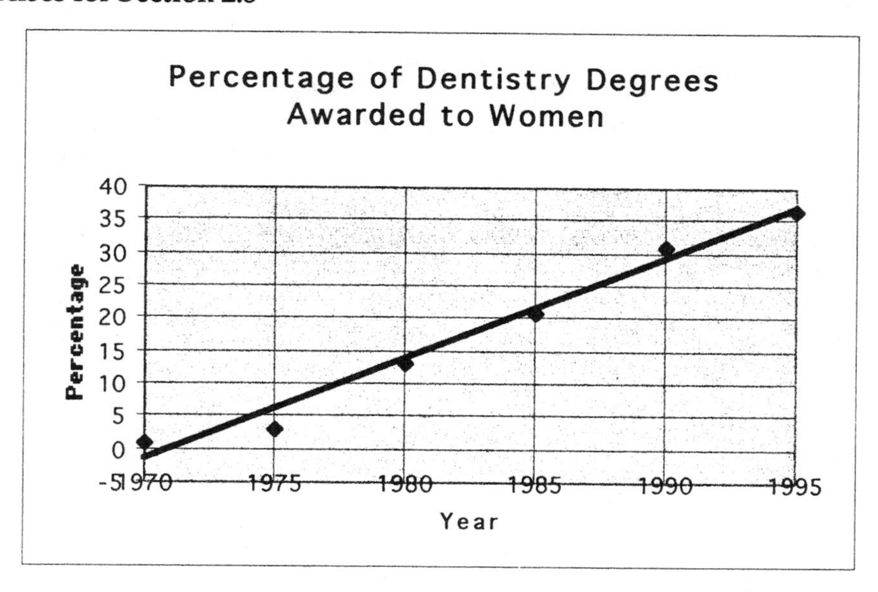

81. a. The graph of the data and the least squares fit line are illustrated above. The equation of the least squares fit is: $y = -1.614 + 1.533x$ (with all decimals rounded to 3 places). Here x = number of years since 1970 and y = percent of dental degrees given to women. The rate of change of percentage of degrees awarded to women per year is approx. 1.5%. Your graphs will probably be different.

b. $100 = -1.614 + 1.533x$ when $x = 101.614/1.533 \approx 66.284$. Thus, if this rate of growth continues, then 100% of dentist degrees would be awarded to women in the year span 2036-37.

c. Growth will slow down. The percent will most likely level off to a ceiling. Your choices of this ceiling will probably differ considerably. Some might even say that it will go back down (possibly in a backlash reaction). There is not enough data to tell.

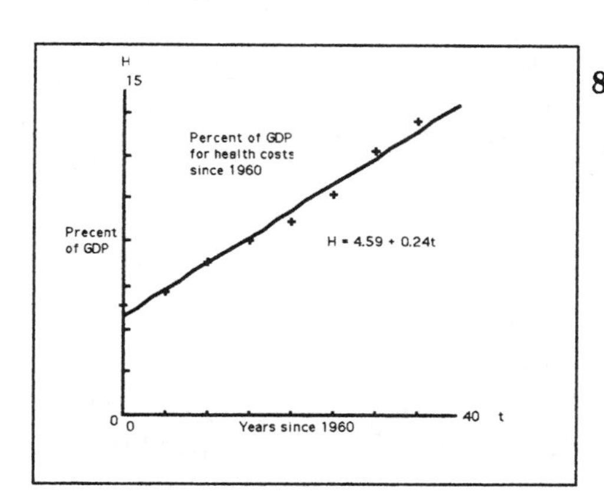

83. a. The graph of the data and the best-fitting straight line derived from technology are illustrated in the graph to the left. The slope of the given line is approximately 0.24 (measured in percentage points per year) and its equation is $H=0.24t+4.59$, where H measure percentage of GDP due to health care costs.

Your eyeball drawn lines will probably be different, as will the corresponding equations. It is expected that you will pick two points in the data set and draw a line through them that is a good approximation to the data and then compute the formula in the usual way. Two such points would be the third and the seventh data points, namely (10, 7.1) and (30, 12.2). These two points define the line whose equation is H = 0.26H + 4.55, which is very close to the best-fitting line provided by technology.

b. Using the best-fitting formula one predicts that in 2010, 50 years from 1960 that health care will be approximately 0.24•50+ 4.59 = 16.6% of the GDP. Your predictions will probably be different but you should plug t = 50 into the formula that was computed. The eyeball formula given in part **a.** gives 0.26•50+4.55 = 17.55%

c. Individual health care costs are going up much more rapidly than the GDP because rises in costs in other areas are much less rapid than in health care.

85. a. The graphs of the data for number of home computers in US, Europe and Japan are found in the diagram below.

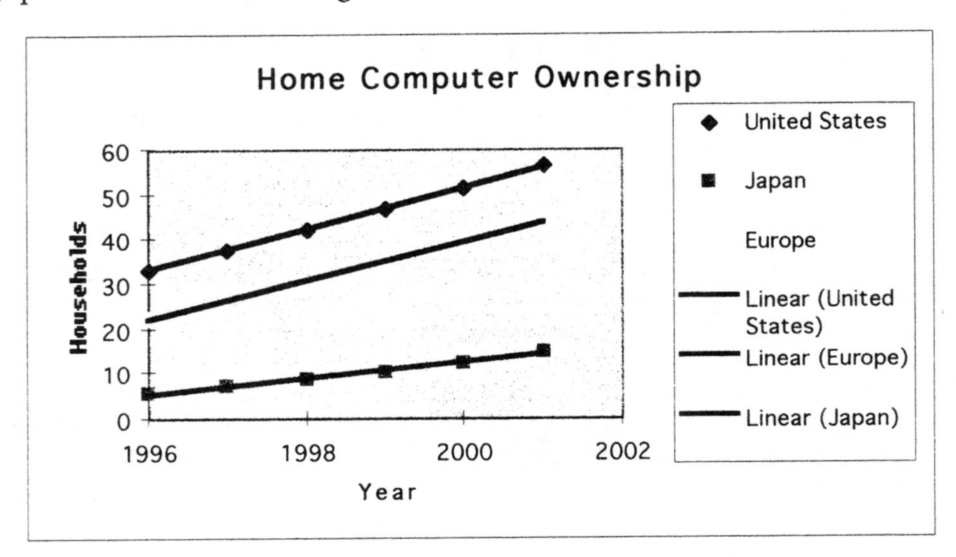

b. The data do suggest that the growth in home computers is roughly linear for each area. The equations for each best-fit line (where x is the number of years since 1996) are:

US:	y = 33.1+ 4.62x
Japan	y = 5.4 + 1.88x
Europe:	y = 22.6 + 4.34x

The graphs of the best-fit straight lines are also in the diagram given above. The slopes represent the overall growth rate of home computers per year. As one can see: the US dominates both in numbers and growth rate. Surprisingly, Japan's numbers and growth rate are very low by comparison. If these rates were to continue, in 2005 the US would have 74.54 million home computers in use, Japan would have 21.55 million and Europe would have 60.86 million.

c. The graphs of the data for each country relevant to the % of houses penetrated by home computers from 1996 through to 2001 are given below.

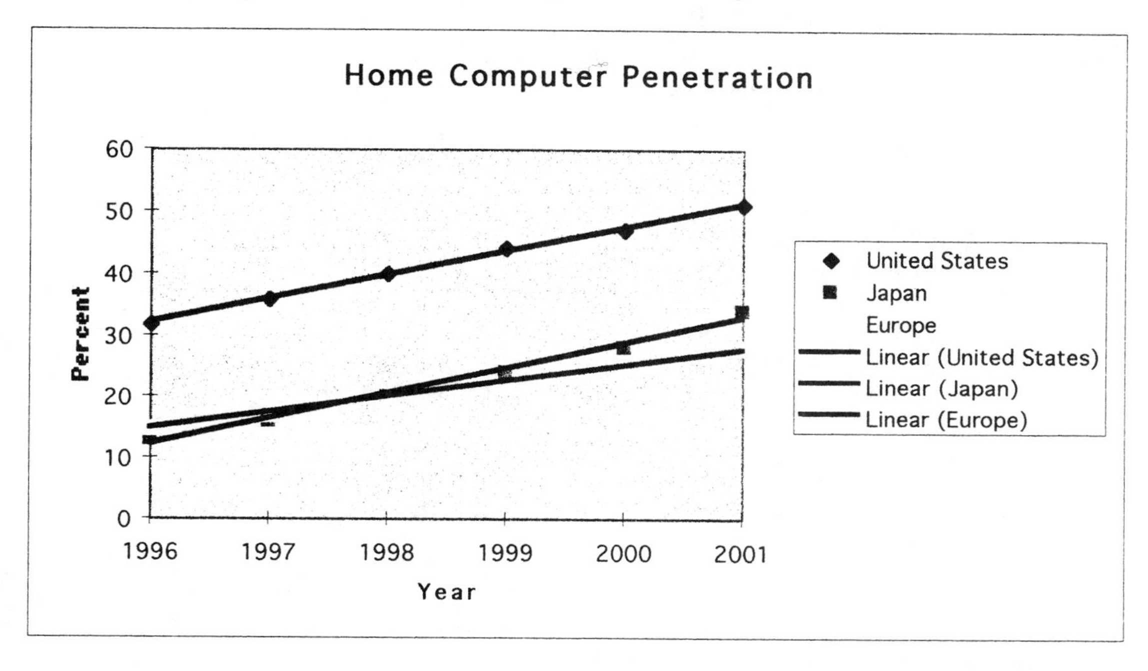

d. Again, the growth seems linear. In the diagram are the plots of the best-fit straight lines for each data set. Their equations are:

US:	$y = 32.238 + 3.771x$
Japan:	$y = 12.143 + 4.143x$
Europe:	$y = 14.667 + 2.600x$

The slopes of these lines represent the overall rate of growth per year in % of households penetrated by home computers. Lastly, if these overall average rates of change should continue, then in 2005, the US would have a 66.177 % penetration, Japan would have a 49.430% penetration and Europe would have a 38.067% penetration.

e. You will probably mention that there has been growth in all three areas and that there is utter dominance by the US in the overall numbers of home computers and percent penetration. But they should also note that Japan (which has the lowest of the 3 in number of computers) is growing in % penetration faster than the US and Europe. The growth rate in the number of personal computers, however, is a very different story. Its growth rate for

Japan is very slow indeed. Nevertheless, the discrepancy in the total number of home computers may be a result of population differences.

87. **a.** The graph of health care cost data as %'s of GDP from 1960 to 1995 is given below. The best straight line fit to this data is also given in the diagram. The equation for this line is: $y = 4.592 + 0.243x$ where x measures years since 1960.

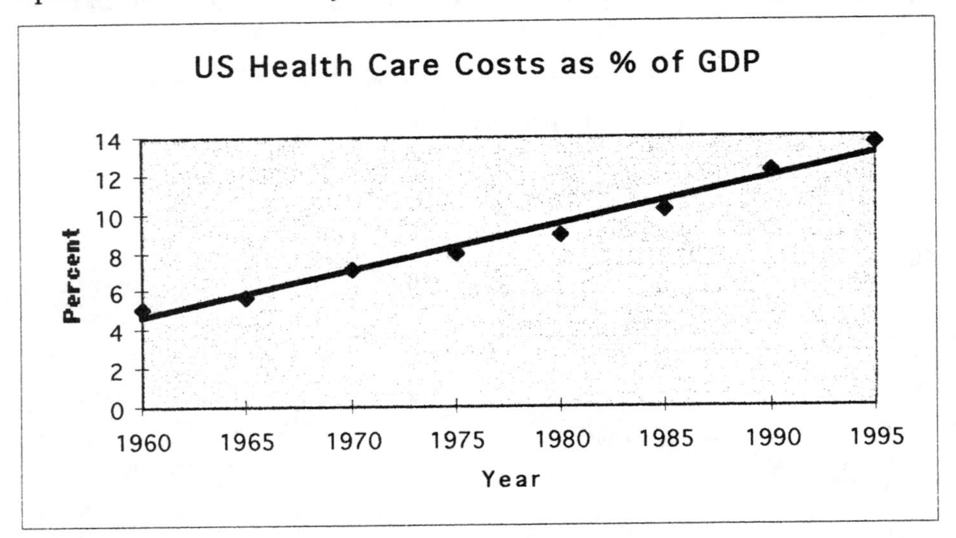

b. In 2010, which is 50 years from 1960, the formula predicts that 16.742% of the GDP will be because of health care costs.

c. Your answers will probably be different. One expected reason is that the GDP has grown very large and thus the percentage for health care is not growing as much as the health care costs themselves.

1. a. 0.65, 0.68, 0.07, 0.70 **b.** 0.07, 0.65, 0.68, 0.70

3. a. The slope of the regression line is 4210; the vertical intercept is -16.520; the correlation coefficient is 0.88.

b. It means that for each additional year of education it is expected that the person's mean personal wages will increase by $4210.

c. $4210, $42,100

5. a. The slope of its graph is 4370

b. It means that the rate of change of mean personal income with respect to years of education is $4370 per year.

c. Mean personal income goes up by $4370 for an increase in education of 1 year; it predicts that mean personal total income rises by $43700 for an increase of 10 years of education.

d. Individual values and outliers are not well described by the regression line.

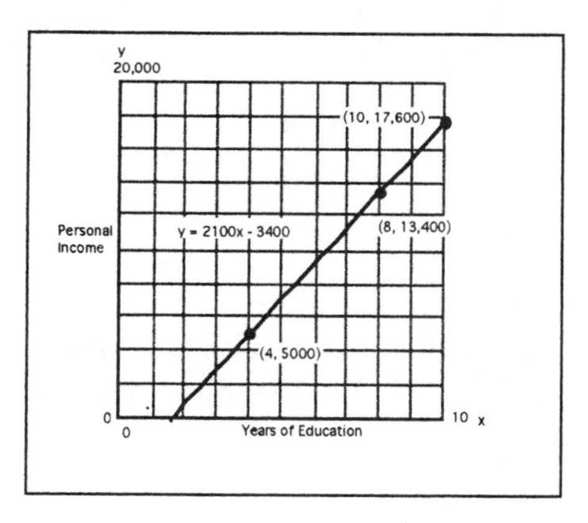

7. a. 2100 in this equation is the rate of change of personal total income with respect to years of education; its units are dollars per year of education.

b. Three points on this line: (4, 5000), (8, 13400) and (10, 17600). The slope, from using the second and third points is:

$$(17600-13400)/(10-8)=4200/2=2100.$$

c. This slope is the same as that given in **a.**

d. The graph is given to the left.

e. A notable difference is the rate at which income increases for each; it is much more rapid for males.

9. a. 0.516 is the rate of change of the mean height of a son in inches, for each increase of 1 inch in height of the father. In other words, if a father is 1 inch taller than another father is, then the mean of height of his son is expected to be on average 0.516 inches more than the mean height of the other father's son.

b. For those fathers who are 64 in. tall, the mean value of their sons' heights is expected to be 66.754 in. For fathers who are 73 inches tall the mean height of their sons is expected to be 71.398 inches.

c. The common height is approximately 70 inches.

d. There are 17 different heights for fathers and for each the mean or average height of sons whose fathers have that height is given. Thus there are only 17 data points plotted.

11. a. and **b.** The best-fit linear equations, (rounded off to the nearest dollar from values obtained using computer software) are:

for private colleges:	$C = 64530 + 819 \bullet yr$
for public colleges:	$C = 1258 + 169 \bullet yr$

where C measures the cost in dollars and yr counts years from 1985.

Below are the graphs of these two linear models along with their respective data sets.

Private Colleges

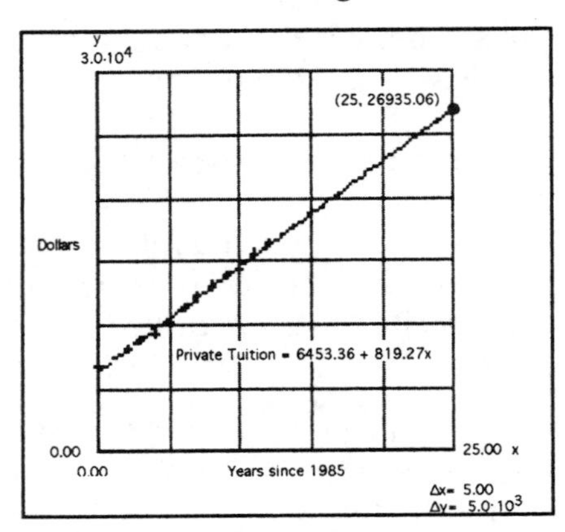

Public Colleges

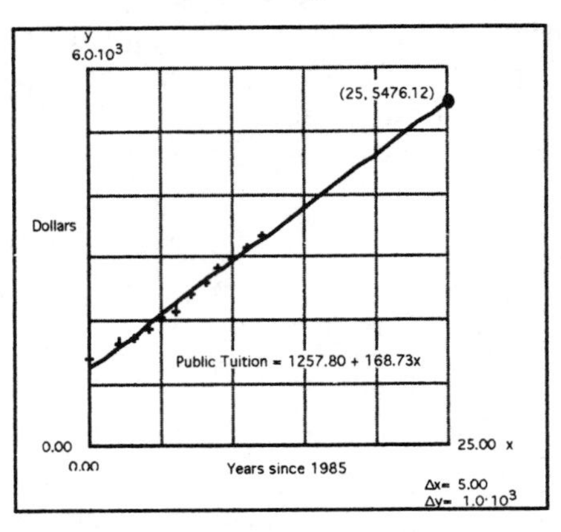

c. Using the rounded off formulae: the estimate for tuition in 2010 for a private college is \$26,928 and for a public college is \$5483. The estimate for a public college seems reasonable but the estimate for a private college seems way too low. Your judgment may differ.

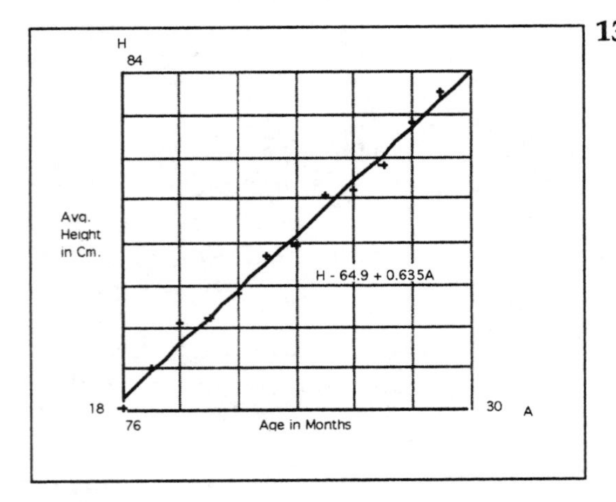

13. a. The equation of the best-fit linear equation is given below. Its graph is given on the left.

$$H = 64.9 + 0.63A$$

b. A represents age in months and H the average height in centimeters. The vertical intercept is 64.9. It represents the (extrapolated) height of a Kalama child at birth. The slope is 0.635; it represents the number of centimeters, on average, that a Kalama child would grow in one month; the correlation coefficient is 0.994 and thus the line is a good fit.

A reasonable domain, built on the data, is from 18 to 29 months.

c. According to the model at 26.5 months a child would be expected to be 81.7 centimeters high.

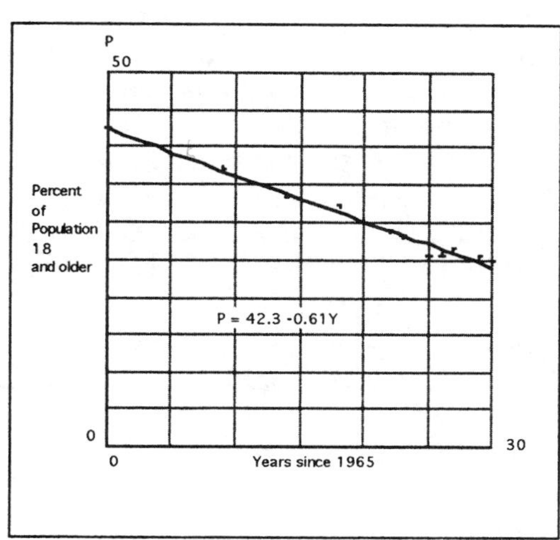

15. a. To the left is a scatter plot of data and a graph of the regression line mentioned in **b.** and **c.** below, where Y = number of years since 1965 and P = % of population 18 and older that smokes.

i. The average rate of change between 1965 and 1995 is (24.7 - 42.4) / (30 - 0) = -0.59; its units are change in % of those who smoke per year.

ii. The average rate of change between 1990 & 1995 is 24.7-25.5)/5 = -0.16. Its units are the same as in **i.**

b. and **c.** Above and to the left is a plot of the regression line along with the data. Its equation is P = 42.35 - 0.61Y where P and Y are defined as in **15. a.** The average rate of change from this line is -0.61% change per year, *i.e.* , the % of smokers is decreasing by 0.61 each year on average. Its correlation coefficient is -0.991825.

Your hand-generated rate may be different -- each will be using his or her eyeball estimate in comparison with the slope of the model's equation.

d. The regression line equation for males is P = 50.33 - 0.83Y and for females it is P = 35.34 - 0.42Y; the correlation coefficient is -0.988 for males and -0.963 for females. In both, Y measures years from 1985 and P is the % of the population being measured who smoke. Note that the starting percentage for males is 18 points higher than the starting percentage for females. Note too that the slope of the male curve is much steeper than that of the female curve.

e. Your answer ought to include the following. From 1985 to 1995: overall, among males and among females, the % of smokers has been decreasing. There was a reversal for a short time period for each group (90-92 overall, 91-92 for males, and 90-92 for females). Year to year, the % of males who smoke is greater than that for females but the rate at which this % has dropped for men is far greater than the rate for females -- so much so that in 1995 the % of males was only 4.4 more than that for females whereas in 1965 it was 18% more. Given the constant negative information being provided about the dangers of smoking, the downward trend will probably continue.

17. One way to check to see if there is hope of a linear relationship is to look at correlation coefficients of each possible pair of data columns. Those close to 1 in absolute value hold promise. Here is a table of these correlation coefficients. The entry in row a, column b is the correlation coefficient with the data in a being the x variable) and the data in b being the y variable. Since the cc is the same no matter which data set is made to be the x values and which made to be the y values, we give that part of the table that is on or below the main diagonal. (Note that the cc. for each data set with itself is always 1.000.)

	Correlation Coefficients for Nations Data Columns								
	GDP	GDP pc	Pers/TV	LifexpM	LifexpF	Pers/Be	Pers/dr	Infmor	Litr %
GDP	1.000								
GDP pc	0.425	1.000							
Pers/TV	-0.085	-0.268	1.000						
LifexpM	0.203	0.607	-0.492	1.000					
LifexpF	0.220	0.662	-0.499	0.947	1.000				
Pers/Bed	-0.105	-0.333	0.304	-0.466	-0.519	1.000			
Pers/Dr	-0.128	-0.383	0.581	-0.623	-0.636	0.286	1.000		
Infmor	-0.213	-0.664	0.498	-0.910	-0.965	0.494	0.662	1.000	
Litr %	0.219	0.591	-0.319	0.724	0.809	-0.479	-0.625	-0.855	1.000

From the table those that show the highest linear correlation are:

Life expectancy of males compared to that for females has a cc. of 0.947.
Life expectancy of males and infant mortality rates has a cc. of -0.910.
Life expectancy of males and literacy % has a cc. of 0.724.
Life expectancy of females and infant mortality rates has a cc. of -0.965.
Life expectancy of females and literacy % has a cc. of 0.809.
Infant mortality and literacy % have a cc. of -0.855.

It would therefore be reasonable to expect linear patterns in their scatter plots. We would not expect it from many of the others.

19. a. and **b.** The scatter plot and regression line for the farm population in absolute numbers are given below on the left. The equation of the regression line is F = -213.21Y + 33458.71, where F stands for the farm population measured in thousands and Y stands for years since 1880. The correlation coefficient is -0.75. This not a very good fit. The scatter plot and regression line for the farm population as a percentage of the whole population is given below on the right. The regression line equation is P = 46.99 - 0.44Y. (Here P measures the percentage of the whole population that were farmers and Y has the same units as above.) Its cc. is -0.989. This is a much better linear fit.

Regression line for 1880-1990 of actual farm population

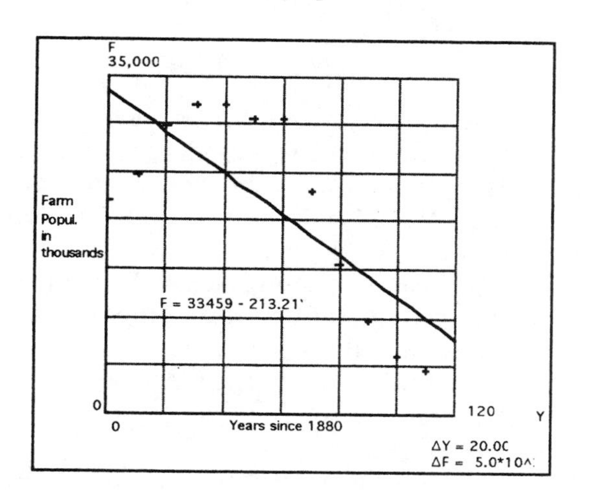

Regression line for 1890-1990 of farm pop. as % of the total population.

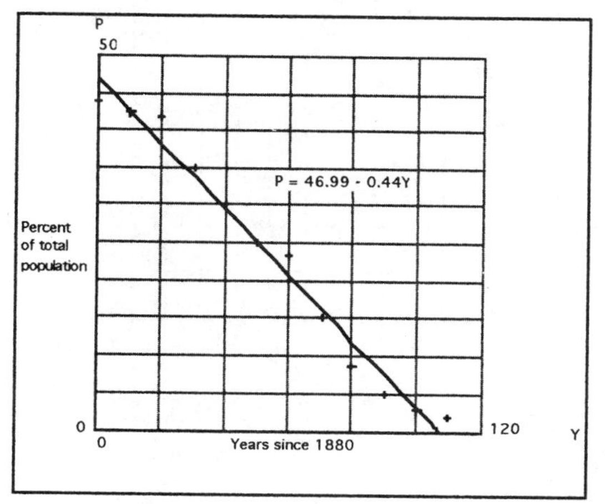

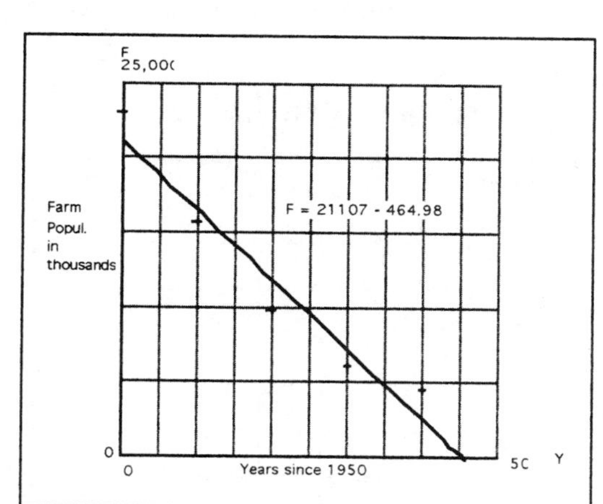

c. If one measures the farm population in absolute numbers from 1950, one gets a much better linear fit as is seen in the graph on the left. The equation for this fit is

$$FP = -464.98x + 21107$$

and the cc. is -0.97.

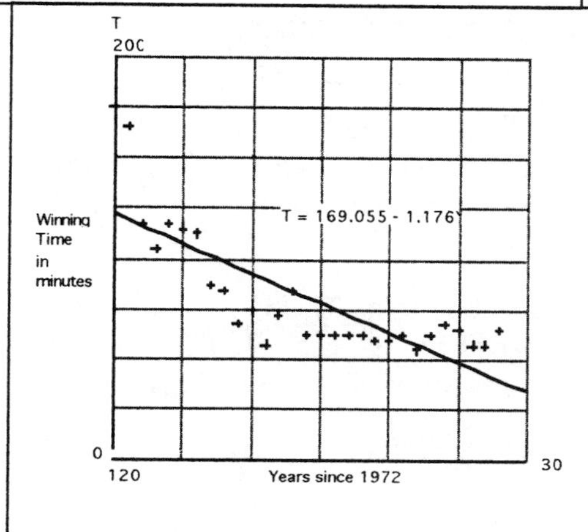

21. a. The best linear fit for the data from 1972 to 2000 is T = 169.055 - 1.176Y, where Y measured years since 1972 and T is measured in minutes. The graph of this data and the best-fit line are given to the left. Its cc. is –0.79. Note the line you draw will probably differ from this one.

b. The predicted winning time for 2010 is approximately 124.372 minutes.

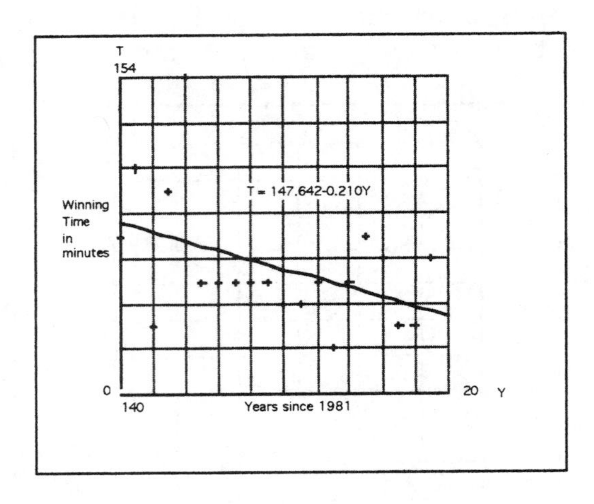

c. The equation of the best-fit line going from 1981 to 2000 is:

$$T = 147.737 - 0.023Y$$

where Y is measured in years since 1981 and T is measured in minutes. The cc is -0.450. The predicted time for 2010 with this model is 141.123 minutes; it is much more realistic than the estimate given in **b.**, even if the cc. is rather low (the scatter diagram with a narrowed range for the winning time tells why the cc. is rather low).

d. Your summary will probably be different but the following might be found. In the early years the winning marathon times for women went steadily down but since 1986 they have leveled off since 1986.

#**23.** Boiling Points in °C

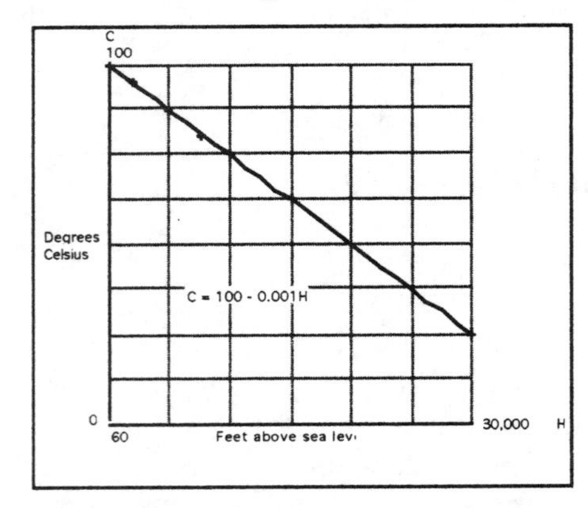

#**23.** Boiling Points in °F

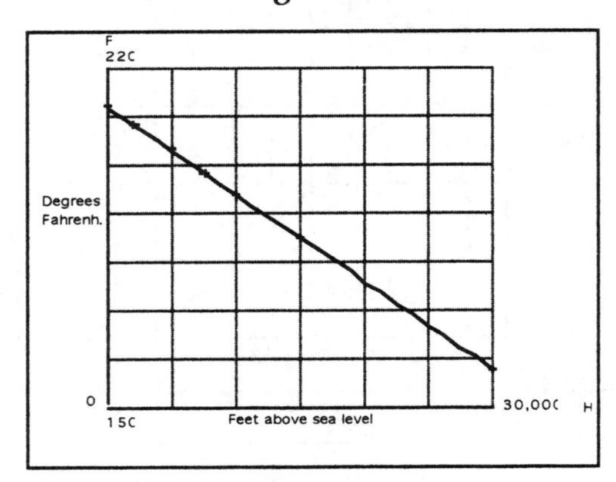

23. a. i. The formula relating °C to altitude is C = 99.91 - 0.00098H or, when suitably rounded off, C = 100 - 0.001H, where H is feet above sea level. The plot of the data and the formula are given above on the left. The correlation coefficient is -0.9999.

ii. The formula relating °F to altitude is F = 211.80 - 0.001793H, or when suitably rounded off, F = 212 - .002H, where H is feet above sea level. The correlation coefficient is -0.9998. It plot is given above on the right.

iii. The answer give here depends on where the you live.

b. On Mt. McKinley water boils at 212 - .002•20320 = 171.36 °F; in Death Valley water boils at 212 - 0.01•(-285) = 212.57 °F.

c. 98.6 = 212 - 0.002•x implies that x = (-113.4)/(-0.002) = 56,700 ft. Thus water in the body would boil at 56,700 ft.

d. 0 = 100 - 0.001•x implies that x = -100/(-0.001) = 100,000 ft. Thus water would boil at °C at 100,000 ft.

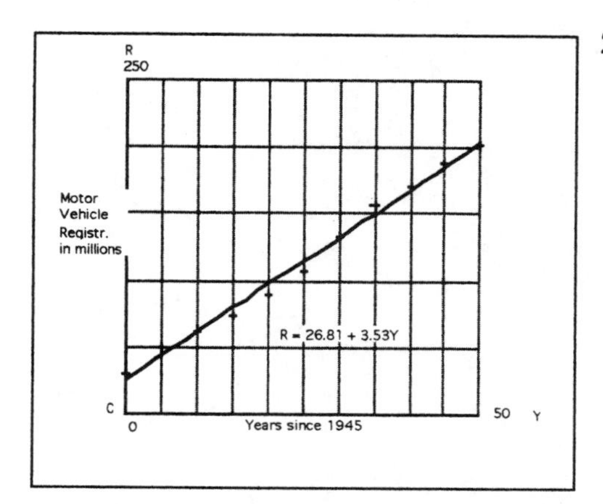

25. a. The graph of the data and the best linear approximation is given in the graph on the left. Its equation is: R = 26.81 + 3.53Y and its cc is 0.9968, where Y stands for years since 1945 and R stands for number of motor vehicle registrations in year Y, measured in millions.

b. Each year, on average the number of motor vehicle registrations increases by 3.53 million.

c. Using the rounded off formula, R = 256.26 million when Y = 65.

27. A high correlation between two factors does not mean that one causes the other. In the case of smoking and lung cancer, however, there is much other evidence to link them causally.

29. In 1993, in a senatorial election in Philadelphia, the Republican candidate received more machine votes than did the Democratic candidate. But the Democratic candidate received so many absentee ballot votes that the combined number of votes (machine and absentee) put him ahead. There were protests that the Democratic candidate unduly influenced absentee ballot votes. The courts threw out the absentee ballots and announced that they would award the election to the Republican if they could be persuaded that his vote count would not have been surpassed by the absentee ballot votes had they been regarded as legitimate.

The Princeton economics professor used regression analysis to show this. He plotted a scatter diagram. The x-axis measured differences between Republican and Democratic machine votes in the last 22 senatorial elections in Philadelphia. The y-axis measured the corresponding differences in absentee ballots. He showed that the points in this scatter diagram tended to fall around a line "representing an ideal correlation" between the machine and absentee tallies. (The newspaper was referring to the regression line of this data.) This ideal situation suggests an agreement between machine and absentee ballot voting. Moreover, the graph shows that the point representing the 1993 election falls well outside the area where 95% of the results would be expected to fall. In fact he showed that the probability of this reversal being due to chance was less than 1%. Based on his arguments, the court awarded the election to the Republican.

<><><><><><><><><><><><><><><><><><><><><><><><><><><><><><><><><><><><><><><><><>

Exercises for Section 3.1

1. "A solution to a system of equations" is a number (or set of numbers) that satisfies all of the equations in the system.

3. a. $4(5) - 3(-10) = 50$ and $2(5) + 2(-10) = -10$ and thus $(5, -10)$ does not solve the given system.

b. The coordinates work for the 1st equation but not the second.

5. a. (40, 3.5)

b. To the left of the point of intersection, the population of Palm City is greater than the population of Johnsonville. To the right of the point of intersection, the population of Johnsonville is greater than that of Palm City.

b. 700 and 150 represent the slopes of the graphs for the equations given; 700 is the rate of change of total cost for gas heat in units of dollars per year since installation; 150 is the rate of change in total cost for solar heat in the same units.

c. 12,000 and 30,000 are the vertical intercepts of the graphs; 12,000 is the initial cost in dollars for installing the gas heating; 30,000 is the initial cost in dollars for installing solar heating.

d. The graphs intersect at approximately $Y = 33$ and $C = 34{,}900$. This means that if gas and solar heating system were installed at the same time, at the prices given, then 33 years later the total cost for gas heat would be the same as the total cost for solar heat and that common cost is approximately $34,900.

e. $Y \approx 32.73$, $C \approx 34909.09$ is a more precise answer. These values have been rounded to 2 decimal places. They are obtained by solving the 2 equations algebraically.

f. From the time of installation up to 32 years, 9 months and 12 days later, the total cost for solar heat is more expensive than that for gas heat and from after that part of the 33rd year the total cost for gas heat becomes greater than the total cost for solar heat.

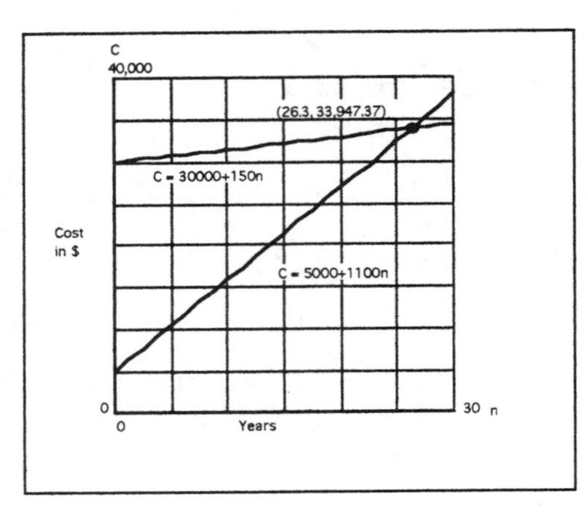

7. a. The graph of the linear system is given on the left with the intersection point marked.

b. 1100 and 159 are the slopes of the heating cost lines; 1100 represents the rate of change of the total cost in dollars for electric heating per year since installation; 150 represents the same thing that it did in **6. b.**

c. 5000 is the initial cost of installing the electric heat in dollars; 30000 has the same meaning as in **6. c.** It cost a lot more initially to install gas than to install electricity.

d. The point of intersection is approximately where $Y = 26$ and $C = 34000$.

e. $Y \approx 26.32$, $C \approx 33947.37$ is a more precise answer; the values have been rounded off to two decimal places; it was obtained using the equations and algebra.

f. Assuming simultaneous installation of both heating systems, the total cost of solar heat was higher than the total cost of electric heat up to year 23 (plus nearly 4 months); after that the total cost of electric heat will be greater than that of solar heat.

[COMMENT: It does not seem to be a realistic assumption that the cost rates will remain unchanged over that long a time period. Did you note this?]

Exercises for Section 3.2

9. a. Set $S_A = S_B$ and get $20000 + 2500n = 25000 + 2000n$ or $500\,n = 5000$ or $n = 10$.

b. The graphs of the two linear equations are given below in the diagram to the left.

From inspecting the graphs it seems that the intersection occurs when $n = 10$ and when the common salary value is 45,000.

Graph for # **9**

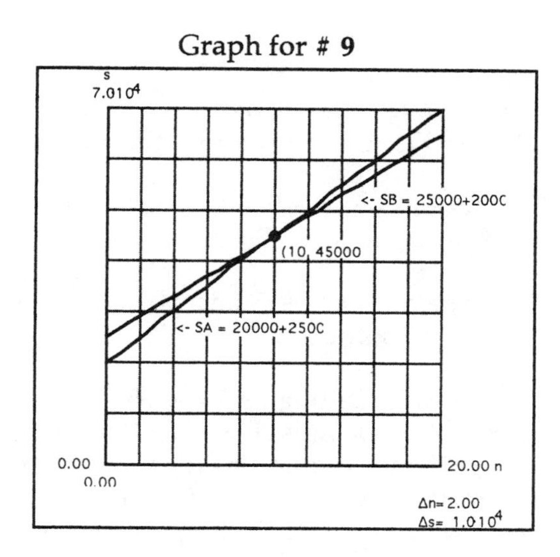

Graph for #**11**

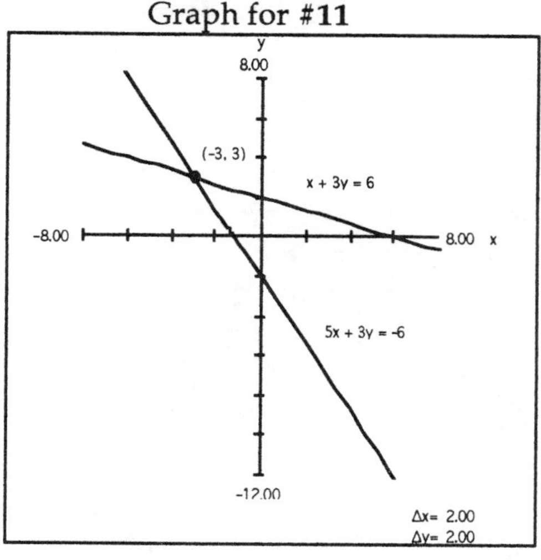

11. a. Subtracting the first equation from the second yields 4x = -12 or x = -3. Putting this value into the first equation gives -3 + 3y = 6 and thus 3y = 9 and y = 3. Putting this value of x into the 2nd equation gives 5(-3) + 3(3) = -6 and thus the solution is (-3,3).

b. The graphs of the two equations and the coordinates of the intersection point are shown above in the figure on the right.

13. Let x = amount to be invested at 4% and let y = amount to be invested at 8%. Then the system of equations to be solved is: x + y = 2000 and .04x +.08y = 100. Substituting y = 2000 - x into the 2nd equation gives, after simplification, that x = $1500 and thus y = $500. [Check: 1500 + 500 = 2000 and .04•1500 + .08•500 =100.]

15. a. Letting y = x - 4/3 from the 2nd equation and substituting this value in the 1st equation we get $\frac{x}{3} + \frac{x - \frac{4}{3}}{2} = 1$. Multiplying both sides by 6 gives 2x +3(x - 4/3) = 6 or 5x -4 = 6 or x = 2 and thus y = 2-4/3 =2/3. [Check: 2-2/3=4/3 and 2/3+1/3= 1.]

b. Substituting y = x/2 from the 2nd equation into the first equation we get x/4 + x/2 = 9 or (3/4)x = 9 or x = 12. Then y = 12/2 = 6. [Check: 12/4 + 6 = 9.]

17. a. The two equations in m and b are: -2 = 2m+b and 13 = -3m+b and the solution is m = -3 and b = 4. [Check: 2(-3)+4 = -2 and -3(-3)+4 =13.]

b. The two equations in m and b are: 38 = 10m+b and -4.5 = 1.5m+b. The solution is m = 5 and b = -12. [Check: 38 = 5•10 – 12 and –4.5 = 5•1.5 - 12]

19. Two equations are equivalent if their graphs are the same, i.e. to say, they have the same sets of solutions. An example is the system 3x = 12 and 2x – 8 = 0.

21. a. 11x + 7y = 68 (4)

b. 9x + 7y = 62 (5)

c. x=3 and y = 5 solve (4) and (5)

d. Thus z = 2•3 + 3•5 -11 = 10

e. Thus the solution is x = 3, y = 5, z = 10 and the check is below:
(1) 2•3 + 3•5 - 10 = 11 (2) 5•3 - 2•5 + 3•10 = 35 (3) 1•3 - 5•5 + 4•10 = 18

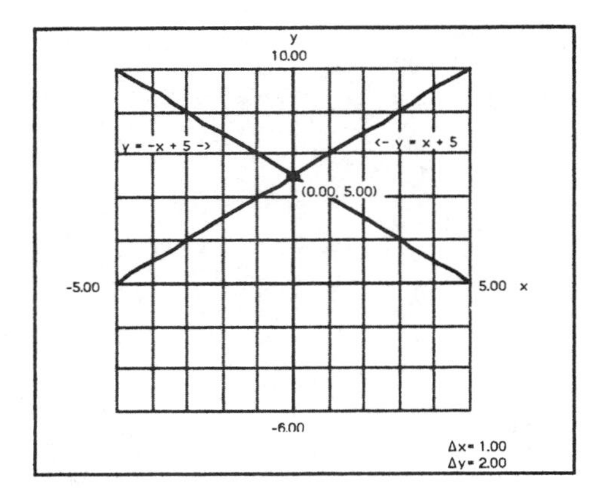

23. a. The system y = x + 5 and y = x + 6 has no solution.

b. The system y = x + 5 and y = -x + 5 has exactly one solution.

c. Algebraically: setting x+5 = -x+5 gives 2x=0 or x = 0 and thus y = 5; alternatively, adding the two equations together gives 2y = 10 or y = 5 and thus x = 0. The graphs of the two lines intersecting at the point claimed is in the diagram on the left.

25. The system of equations has no solution if the graphs of the two equations are parallel and distinct lines. This occurs when $m_1 = m_2$ (parallel means same slope) and $b_1 \neq b_2$ (different vertical intercepts).

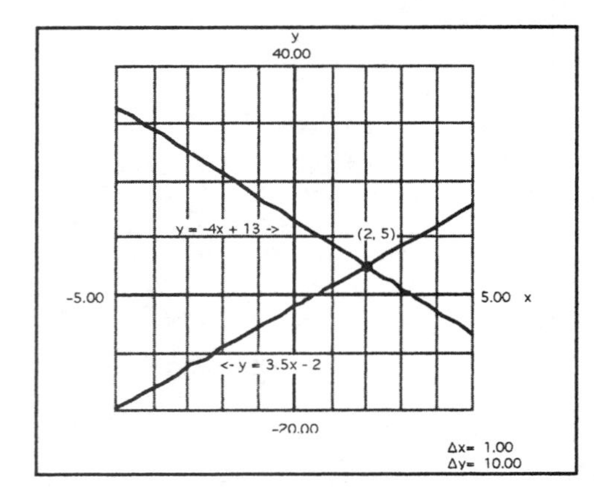

27. a. The equations of this pair of lines are:

y - 5 = -4(x - 2) and y - 5 = 3.5(x - 2)

or in simplified form:

y = -4x + 13 and y = 3.5x - 2

b. The graphs of the two equations are given in the diagram on the left.

[Check: -4•2+13 = 5 and 3.5•2 -2 = 5]

29. a. (A + B)T = Q, where T = number of hours and A and B are measured in bricks per hour. Thus it will take T = Q/ (A+B) hours to lay A+B bricks.

b. Let x count the hours Alpha works. Then x - 2 is the number of hours that Beta works. Thus an equation that represents the time when the have both laid the same number of bricks is: 50x = 70(x - 2) = 70x - 140 or 20x = 140 or x = 7 hours.

c. Here is the table asked for:

A time (hrs.)	A bricks laid	A cost ($)	B time (hrs.)	B bricks laid	B cost ($)
1	50	42	1	70	45
2	100	72	2	140	90
3	150	102	3	210	135
4	200	132	4	280	180
5	250	162	5	350	225
6	300	192	6	420	270
7	350	222	7	490	315
8	400	252	8	560	360

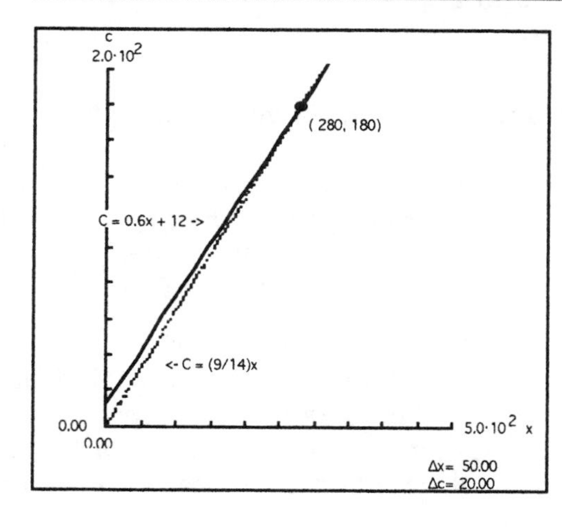

The equation for the cost of Alpha laying bricks vs. the number of bricks laid is given by a straight line that goes through (50, 42) and (100, 72) and thus the slope of this line is (72 - 42)/(100 - 50) = 30/50 = 0.60 and thus the equation is:

$$C - 72 = 0.60(x - 100) \text{ or } C = 0.60x + 12.$$

The equation for the cost of Beta laying bricks vs. the number of bricks laid is given by the straight line that goes through (70, 45) and (140, 90). Thus the slope of this line is (90 - 45)/(140 - 70) = 45 / 70 = 9 / 14 and the equation is $C - 90 = (9/14)\bullet(x - 140)$ or $C = (9/14)x$. From the graph we estimate that the point at which Alpha and Beta have laid the same number of bricks and have been paid the same amount of money when x = 280 bricks and C = $180.

It is cheaper to hire Alpha when there are more than 280 bricks to be laid.

d. Alpha lays 50 bricks an hour and thus $A_{bricks} = 50T_\alpha$, where T_α is measured in hours and A in bricks. Similarly Beta lays 70 bricks and hour and thus $B_{bricks} = 70T_\beta$. Also, $A_{cost} = 12 + 30T_\alpha$ is cost of Alpha working T_α hours; $B_{cost} = 45T_\beta$ is the cost of Beta working T_β hours.

Alpha and Beta lay the same number of bricks and get paid the same amount when the number of bricks is 280 and the payment is $180. But they have worked a different number of hours: Alpha lays 280 bricks in 5.6 hours or 5 hours and 36 minutes. and Beta lays 280 bricks in 4 hours.

31. a. $B_y = 30$, $A_y = 0.625A_x$

b. The common point in space that both planes will eventually occupy is where $x = 48$ and $y = 30$. (It is the intersection point of the graphs of $y = 30$ and $y = 0.625x$. These are equations for constant altitude of the flight paths of the two planes.)

c. B, in going from (-30, 30) to (48, 30), travels a distance of 78 miles and this takes B 13 minutes to do (since it is traveling at 6 miles/minute). A, in traveling from (80, 50) to (48, 30), takes $[(30\text{-}50)^2 + (48 - 80)^2]^{0.5} = [400 + 1024]^{0.5} = 1424^{0.5} =$ 37.736 miles and this will take 18.868 minutes (since plane B is traveling at 2 miles per minute). Thus plane A will arrive at this point nearly 6 minutes after plane B. It is, therefore, a safe situation.

Exercises for Section 3.3

pay a higher price. This shift is illustrated in the graph above on the right.

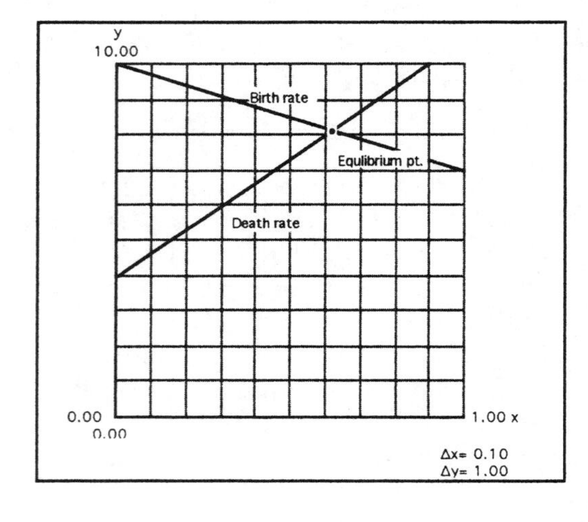

33. a. To the left is a rough linear sketch of birth rate and death rate over population density. The point of intersection or equilibrium point is marked.

b. The growth rate of population at the equilibrium point is zero: since the population growth rate = birth rate - death rate and the latter two are equal in value.

c. If the death rate decreases (and the birth rate stays the same) the equilibrium point tends to shift down the right. (See the graph to the left below.)

d. If the death rate increases (and the birth rate stays the same) the equilibrium point tends to shift right and go up. (See the graph to the right below.)

#33. c. Lower death rate

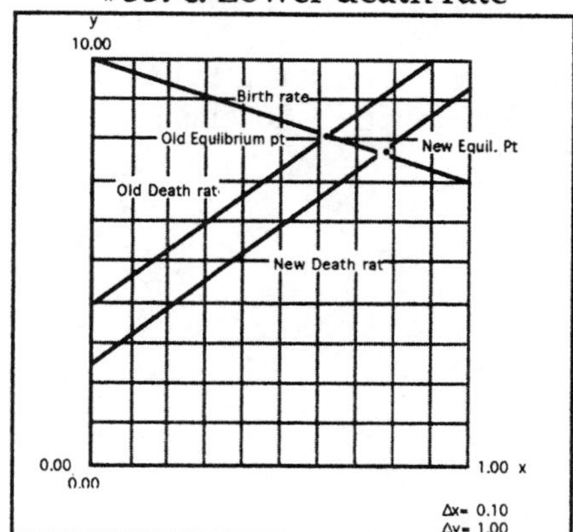

#33. d. Higher death rate:

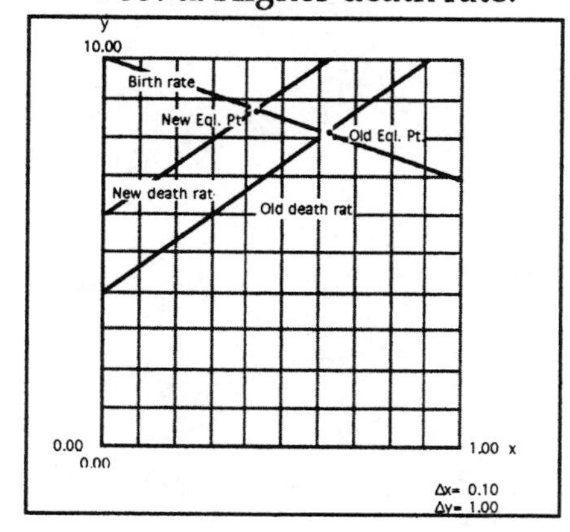

35. Answers will vary from state to state. Check your answer against your local tax form itself.

37. a. $y = 1$ for $0 \le x \le 1$
$= x$ for $x > 1$

b. $y = 1 - x$ for $0 \le x \le 1$
$= 1.5x - 1.5$ for $x > 1$.

39. If $f(x) = 2x + 1$ for $x \le 0$
$= 3x$ if $x > 0$

then $f(-10) = -19$, $f(-2) = -3$, $f(0) = 1$, $f(2) = 6$, and $f(4) = 12$.

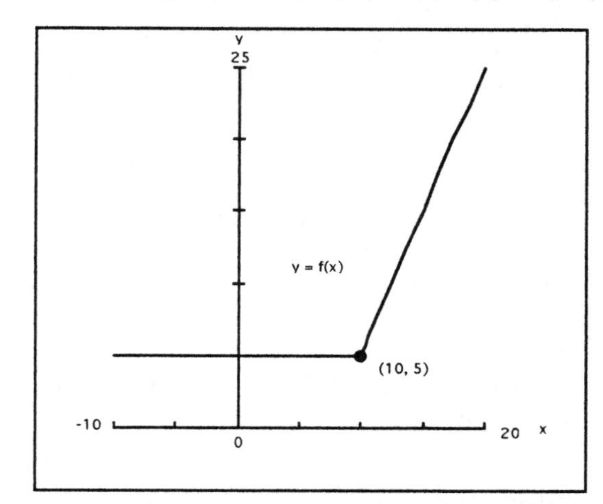

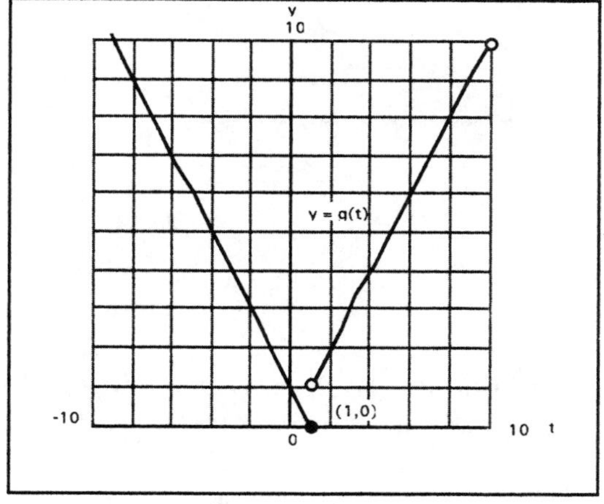

41. a. Table for f whose domain is all real x (graph is above)

x	f(x)
0	5
3	5
8	5
10	5
15	15
20	25

b. Table for g whose domain is $-10 \le t < 10$ (graph is above)

t	g(t)
-10	11
-5	6
1	0
3	3
5	5
10	10

43. a. flat tax: $f(x) = 0.10 \bullet x$ if $x > 0$

grad tax $g(x) = 1000$ if $0 < x \leq 20000$;
$= 1000+0.20(x-20000) = 0.2x - 3000$ if $x > 20000$

b. Here is a small table:

Income	Flat	Graduated
0	0	0
10000	1000	1000
20000	2000	1000
30000	3000	3000
40000	4000	5000

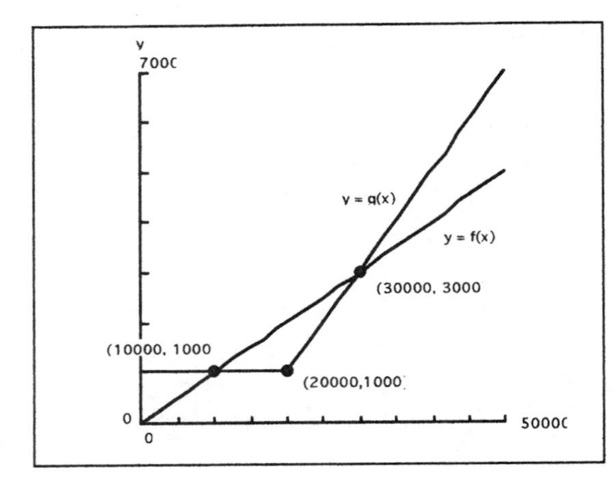

c. The graphs of both tax plans and their intersection points are given on the left. Note that the intersection points that are marked are given in the table.

d. We should try to solve for intersection points with each part of the formula for g. Thus:

$$0.10x = 1000 \text{ gives } x=10{,}000$$

and

$$1000+0.20(x-20000) = 0.10x$$

gives

$$0.10x = 3000 \text{ or } x = 3000.$$

e. The flat tax is more than the graduated tax for $10{,}000 < x < 30{,}000$. The graduated tax is more for $0 < x < 10{,}000$ and $x > 30{,}000$.

f. If $0 < x < 1000$, then the tax under the graduated plan would be \$1000; thus there would be negative income.

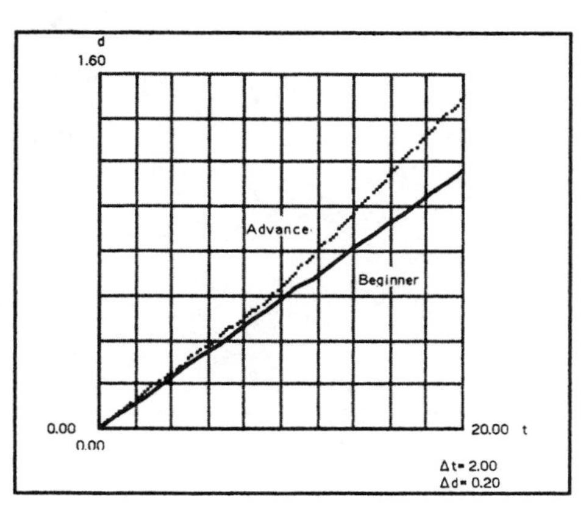

45. The graphs for **a.** and **b.** are found on the left.

a. For $0 \le T \le 20$: $D_{beginner} = (3.5/60)T$ since there is 1/60 of an hour in a minute; note that T is measured in minutes and $D_{beginner}$ is measured in miles.

b. For $0 \le T \le 10$:

$$D_{advanced} = (3.75/60)T$$

and for $10 < T \le 20$ we have that

$$D_{advanced} = 0.625 + (5.25/60)(T\text{-}10)$$
$$= 0.0875T - 0.25$$

47. a. C=7 if N<100; C=7 if N=100; C = 3.5 if N>100; here cost is measured in cents.

b. $C(N) = 7$ if $0 < N \le 100$; $C(N) = 3.5$ if $N > 100$; again cost is measured in cents.

c. C(N) is the charge per photocopy if N copies are to be made.

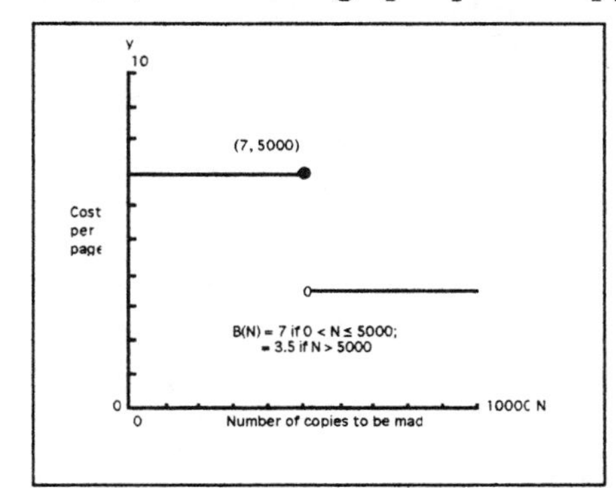

d. The function B is graphed on the left; the cost is measured in cents and N counts the number of pages copied.

$$B(N) = 7 \text{ if } 0 < N \le 5000,$$
$$= 3.5 \text{ if } N > 5000$$

e. $TC(N) = 0.07N$ if $0 < N \le 100$
$= 7 + 0.035(N\text{-}100)$ if $N > 100$

The cost is now measured in dollars and N represents the number of pages copied.

f. $TB(N) = 0.07N$ if $0 < N \le 5000$,
$= 350 + 0.035(N\text{-}5000)$ if $N > 5000$

The cost is measured in dollars and N represents the number of pages copied.

Exercises for Section 4.1

1. a. 10^6 **b.** 10^{-5} **c.** 10^9 **d.** 10^{-3}

3. a. 10^{-2} m **b.** $4 \bullet 10^3$ m **c.** $3 \bullet 10^{12}$ m **d.** $6 \bullet 10^{-9}$ m

5. a. $2.9 \bullet 10^{-4}$ **b.** $6.54456 \bullet 10^2$ **c.** $7.2 \bullet 10^5$ **d.** $1.0 \bullet 10^{-11}$

7. a. 723,000 **b.** 0.000526 **c.** 0.001 **d.** 1,500,000

9. a. 9 **b.** 9 **c.** 1000 **d.** -1000

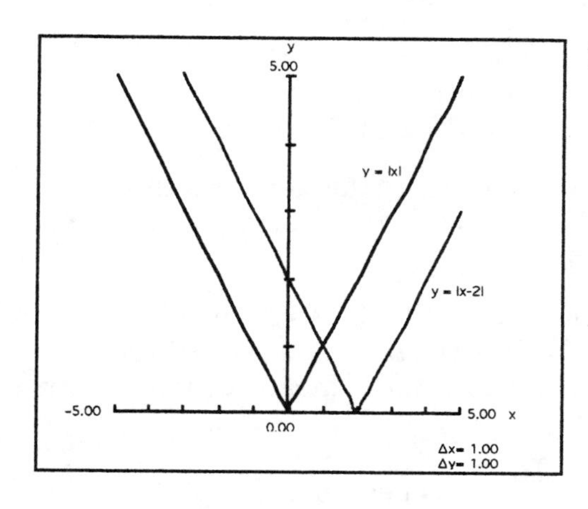

11. a. Here is a small table of values for $y = |x|$

x	|x|
-2	2
-1	1
0	0
1	1
2	2

b. The graphs of $y = |x|$ and $y = |x-2|$ are displayed on the left to allow for comparison.

Exercises for Section 4.2

13. a. 10^7 **b.** $1.1 \bullet 10^4$ **c.** $2 \bullet 10^3$ **d.** x^{15} **e.** x^{50}
f. $1.6409 \bullet 10^4$ (scientific notation was chosen; other forms are also allowed).
g. z^5 **h.** 1 or as is **i.** 3^{-1} **j.** 4^{11}

15. a. $16a^4$ **b.** $-2a^4$ **c.** $-x^{15}$ **d.** $-8a^3 b^6$
e. $32x^{20}$ **f.** $18x^6$ **g.** $2500a^{20}$ **h.** as is -- nothing is simpler

17. a. $-\left(\frac{5}{8}\right)^2 = -\frac{25}{64}$ **b.** $\left(\frac{3x^3}{5y^2}\right)^3 = \frac{3^3 x^9}{5^3 y^6} = \frac{27x^9}{125y^6}$

c. $\left(\frac{-10x^5}{2b^2}\right)^4 = \frac{10^4 x^{20}}{4^2 b^8} = \frac{10{,}000x^{20}}{16b^8} = 625\frac{x^{20}}{b^8}$ **d.** $\left(\frac{-x^5}{x^2}\right)^3 = -x^9$

19. a. $10^9/10^6 = 10^3 = 1000$ **b.** $1000/10 = 10^2 = 100$
c. $1000/0.001 = 10^6 = 1{,}000{,}000$ **d.** $10^{-6}/10^{-9} = 10^3 = 1000$

21. a. Japan's population density = $(126.1 \bullet 10^6)/(152.5 \bullet 10^3) = 826.89$ persons per square mile.

b. US population density = $(272.6 \bullet 10^6)/(3620 \bullet 10^3) = 75.30$ persons per square mile.

c. The population density of Japan is almost 11 times as great as that for the US.

23. a. If $a = 0$, then $(-a)^n = 0$. If $a \neq 0$, then when n is an even integer, $(-a)^n$ is positive and when n is odd, then $(-a)^n$ has the opposite sign of a.

b. This is answered in part **a**.

25. Three cases are distinguished:

If $n = 0$, then $(ab)^0 = 1$ and $a^0 \bullet b^0 = 1 \bullet 1 = 1$

If $n > 0$, then $a^n = a \bullet \cdots \bullet a$ (n factors) and $b^n = b \bullet \cdots \bullet b$ (n factors) and thus $a^n \bullet b^n = (a \bullet \cdots \bullet a) \bullet (b \bullet \cdots \bullet b) = (ab) \bullet \cdots \bullet (ab)$ (n factors), after rearrangement, and this product is what is meant by $(ab)^n$ when $n > 0$.

If $n < 0$, then $a^n = (1/a) \bullet \cdots \bullet (1/a)$ (-n factors) and $b^n = (1/b) \bullet \cdots \bullet (1/b)$ (-n factors) and thus $a^n \bullet b^n = (1/a) \bullet \cdots \bullet (1/a) \bullet (1/b) \bullet \cdots \bullet (1/b) = (1/ab) \bullet \cdots \bullet (1/ab)$ (-n factors), after rearrangement, and this product is what is meant by $(ab)^n$ when $n < 0$.

27. a. For UK : $(10.1 \bullet 10^9)/(59 \bullet 10^6) = 171.186 = 1.71 \bullet 10^2$ kwhs per person;
$(10.1 \bullet 10^9)/(9.4525 \bullet 10^4) = 1.069 \bullet 10^5$ kwhs per square mile.

b. For US: $(55.6 \bullet 10^9)/(2.7 \bullet 10^8) = 2.059 \bullet 10^2$ kwhs per person
$(55.6 \bullet 10^9)/(3675031) = 1.5129 \bullet 10^4$ kwhs per square mile.

c. $(1.069 \bullet 10^5)/(15.129 \bullet 10^3) = 7.0659$. Thus UK is generating per square mile 7 times the kwhs that US is.

d. Since $(2.059 \bullet 10^2)/(171.186) = 1.20$ we see that the US is generating 1.2 times the number of kwhs per person that the UK is generating.

Exercises for Section 4.3

29. a. $\frac{1}{2^2 x^4} = \frac{1}{4x^4}$ **b.** xy^{12} **c.** $\frac{1}{x^3 y^2}$ **d.** $(x+y)^{11}$ **e.** $\frac{ab^7}{c^6}$

[COMMENT: "simplify" is taken to mean "put the fraction in lowest terms and make all exponents positive".]

31. a. $4.6 \bullet 10^{10}$ **b.** $4.07 \bullet 10^3$ **c.** $1.525 \bullet 10^{11}$
d. $5.1669 \bullet 10^{-8}$ **e.** $2.3833 \bullet 10^{158}$ **f.** $2.601 \bullet 10^{-21}$

33. 200 times longer or $1.6 \bullet 10^{-2}$ seconds.

35. a. $\left(\frac{1}{x^2}-\frac{1}{y}\right)\bullet(xy^2)=\left(\frac{y-x^2}{x^2y}\right)xy^2=(y-x^2)\bullet\frac{y}{x}=\frac{y^2-x^2y}{x}$ (this one is difficult)

b. $\frac{x^4y^6}{5^2}=\frac{x^4y^6}{25}$

Exercises for Section 4.4

37. a. Your estimate might be different. There are approximately 30.77 cm. per foot.
b. Your estimate might be different. 1 m = 3.28 ft. and thus 1 ft = (1/3.28) =0.305 m.

39. 1m = 3.28 ft and thus 9.8 mps = 9.8•3.28 = 32.144 ftps

41. a. 186,000 miles per second or one hundred and eighty six thousand miles per second

b. $9.4608 \bullet 10^{15}$ meters per year

43. 500 seconds; 500 seconds or 8 minutes and 20 seconds from now.

45. Light travels 186000•5280 ft. per sec. and thus it travels $186000\bullet5280/10^9 \approx 0.98208$ ft per nanosecond.

47. a. Clinton's BMI ≈ $(216/2.2)/(74/39.37)^2 \approx 27.79$ kg/m^2; overweight.
b. BMI in lbs and inches is BMI = $(\text{lbs}/2.2)/(\text{inches}/39.37)^2 = [(39.7)^2/2.2](\text{lbs}/\text{in}^2)$ or 704.5 lbs/in^2 and thus for Clinton we get: $704.5\bullet216/74^2 \approx 27.79$
c. Your estimate might be different but note that 0.45 ≈ 1/2.2 and .0254 ≈ 1/39.37
d. A kilogram, more precisely, is 2.2046 lbs and $39.37^2/2.2046 \approx 703.07$ and thus Brent Kigner is correct.

49. a. 10^{-35} m $= 10^{-35}\bullet10^{-3} = 10^{-38}$ km
b. 10^{-35} m $= 10^{-35}\bullet0.00062 = 6.2\bullet10^{-39}$ miles
c. $x = 10^{-35}/3\bullet10^{-8} = 3.33\bullet10^{-28}$

Exercises for Section 4.5

51. a. $\sqrt{\frac{a^2b^4}{c^6}}=\frac{ab^2}{c^3}$ **b.** $\sqrt{36x^4y}=6x^2\sqrt{y}$ **c.** $\sqrt{\frac{49x}{y^6}}=\frac{7\sqrt{x}}{y^3}$ **d.** $\sqrt{\frac{x^4y^2}{100z^6}}=\frac{x^2y}{10z^3}$

[COMMENT: All these answers assume that a, b, c, x, y, z are positive.]

53. $r=\sqrt{\frac{20}{4\pi}}\approx1.262$ meters

55. a. 0.1 **b.** 1/5 **c.** 4/3 **d.** 0.1

57. a. 4 **b.** -4 **c.** $\sqrt{2} \approx 1.414$ **d.** not defined **e.** $2\sqrt{2} \approx 2.8284$ **f.** 1

59. a. 9 **b.** 8 **c.** $\frac{1}{125}$ **d.** $(1/3)^3 = 1/27$

61. $V = (4/3)\pi r^3$ and thus if $V = 4$ then $r^3 = 3/\pi$. If one uses 3 as a crude estimate of π then r is approximately 1 foot. (A more precise estimate, from using a calculator, is 0.985 ft.)

63. a. 3673.01 lbs. **b.** 6344.62 lbs.

Exercises for Section 4.6

65. a. $(50\bullet10^3)\bullet(3\bullet10^8)\bullet(365\bullet24\bullet60^2) = 4.7304\bullet10^{20}$ = number of meters in 50,000 light years. Hence, the Milky Way is $(4.7302\bullet10^{20})/(0.5\bullet10^{-4}) = 9.4608\bullet10^{24}$ times larger than the first life form. Thus the Milky Way is nearly 25 orders of magnitude larger than the first living organism on Earth.

b. $(100\bullet10^6)/(100\bullet10^3) = 10^3$; thus Pleiades is 3 orders of magnitude older than *homo sapiens*.

67. a. $2.0\bullet10^{-5}$ in.
b. $2.0\bullet10^{-5}/39 = 5.128\bullet10^{-7}$ meters
c. The name is a bit off -- by 2 orders of magnitude (looking at meters).
d. The tweezers would have to be made able to grasp things 2 orders of magnitude smaller than they can grasp now.

69. a. **i.** radius of the moon = 1,758,288.293 meters or $1.76\bullet10^6$ meters.
ii. radius of Earth = 6,400,000 meters or $6.4\bullet10^6$ meters.
iii. radius of the sun = 695,414,634.100 meters or $6.95\bullet10^8$ meters.

b. **i.** The surface area of a sphere is $4\pi r^2$ and $[3906/1092.27]^2 = 12.79$. Thus the surface area of Earth is one order of magnitude bigger.
ii. The volume of a sphere is $(4/3)\pi r^3$ and $[432000/1092.67]^3 = 6.18\bullet10^7$ and thus the volume of the sun is 7 orders of magnitude bigger than the volume of the moon.

71. Scale a is additive scale while scale b appears to be logarithmic.

73. a. Being 5 orders of magnitude larger than the first atoms means that it is 10^{-5} m and thus would appear at -5 on the log scale.
b. Being 20 orders of magnitude smaller than the radius of the sun means that it is 10^{-11} m and thus would appear at -11 on the log scale.

75. Assuming that both scales start at 10^0 and each division is a power of 10 more than the previous one, then since log(708) = 2.850 and log(25) = 1.397 the point would be about 85% of the way between 10^2 and 10^3 along the x axis and about 40% of the way between 10^1 and 10^2 along the y axis.

77. a. 10^{-7} cm

b. A radio wave is 10^5 cm long on average

c. 13 orders of magnitude (X-rays, on average, are 10^{-8} cm. long.)

d. From 10 Å to 10^3 Å

e. From 10^4 Å to 10^7 Å

Exercises for Section 4.7

79. a. log(10,000) = 4 **b.** log(0.01) = -2 **c.** log(1) = 0 **d.** log(0.00001) = -5

81. Since log(375) ≈ 2.574 we have that 375 ≈ $10^{2.574}$

83. a. $\log_{10}(100) = 2$ **b.** $\log_{10}(10{,}000{,}000) = 7$ **c.** $\log_{10}(0.001) = -3$
d. $10^1 = 10$ **e.** $10^4 = 10{,}000$ **f.** $10^{-4} = 0.0001$

85. a. 1 < log 11 < 2 (since 10 < 11 < 100 and log 11 ≈ 1.0414)
b. 4 < log 12000 < 5 (since 10,000 < 12,000 < 100,000 and log 12,000 ≈ 4.0792)
c. -1 < log 0.125 < 0 (since 0.1 < 0.125 < 1 and log 0.125 ≈ -0.9031)

87. a. Multiply by 10^{-3} **b.** Multiplying by $\sqrt{10}$

c. Multiplying by 10^2 **d.** Multiplying by 10^{10}.

89. a. pH = $-\log(10^{-7}) = 7$

b. pH $= -\log(1.4 \bullet 10^{-3}) = 3 - \log(1.4) \approx 2.85$

c. $11.5 = -\log([H+])$; thus $[H+] = 10^{-11.5} \approx 3.16 \bullet 10^{-12}$

d. A higher pH means a lower hydrogen ion concentration because y = -log(x) is a decreasing function.

e. Pure water is neutral, orange juice is acidic and ammonia is basic. In plotting, one uses the top numbers to find the right spots. Thus water would be placed at the 7 mark, orange juice 85% of the way between the 2 and 3 marks and ammonia half way between the 11 and 12 marks.

Exercises for Section 5.1

1. a. Initial population = 275; growth factor = 3
b. Initial population = 15,000; growth factor = 1.04
c. Initial population = $6 \bullet 10^8$; growth factor = 5

3. a. $G(t) = 5.10^3 \bullet 1.185^t$ cells/ml **b.** $G(8) = 5.10^3 \bullet 1.185^8 = 19440.92$ cells/ml

5. a. and **b.**

t	Q(t) = 5+1.5t	Avg. Rate of Ch. between t-1 and t	Q(t) = $5 \bullet 1.5^t$	Avg. Rate of Ch. between t-1 and t
0	5.0	n.a.	5.0	n.a.
1	6.5	1.5	7.5	2.5
2	8.0	1.5	11.25	3.75
3	9.5	1.5	16.875	5.625
4	11.0	1.5	25.313	8.4375

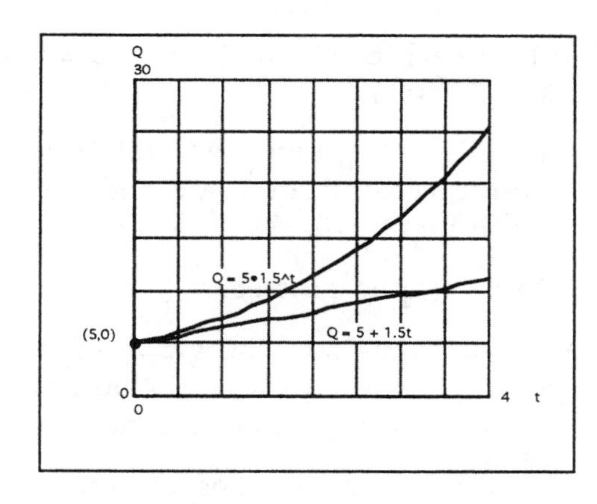

c. The graphs of the two functions are in the diagram to the left. They intersect only at (0, 5)

From the shapes of the curves, for $t \geq 0$, the linear function's graph will never lie above the exponential's graph.

The exponential will grow rapidly for values of $t > 4$ and its Q values will stay much bigger in value than the linear function's values.

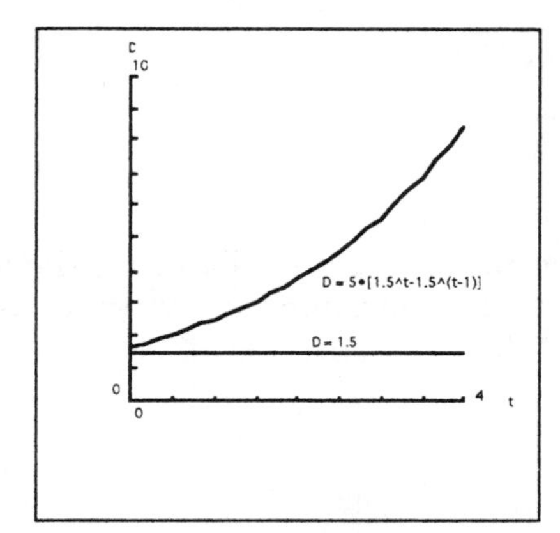

d. The plots of the two average rate of change functions are given in the diagram on the left.

The rate of change for the exponential is ever increasing while that for the linear function is, as expected, a constant.

Solutions to the Odd-Numbered Exercises in Chapter 5

Exercises for Section 5.2

7. a. $f(t) = 10{,}000 \bullet 0.4^t$ **b.** $g(T) = 2.7 \bullet 1013 \bullet 0.27^T$ **c.** $h(x) = 219 \bullet 0.1^x$

9. a. This is not exponential since the base is the variable.

b. This is exponential; the decay factor is 0.5; the vertical intercept is 100.

c. This is exponential; the decay factor is 0.999; the vertical intercept is 1000.

11. a. $P(t) = 100 \bullet b^t$; $99.2 = 100 \bullet b$ or $b = 0.992$ and therefore $P(t) = 100 \bullet 0.992^t$

b. $P(5)) = 100 \bullet 0.992^{50} \approx 66.9$ grams; $P(500) = 100 \bullet 0.992^{500} \approx 1.8$ grams.

Exercises for Section 5.3

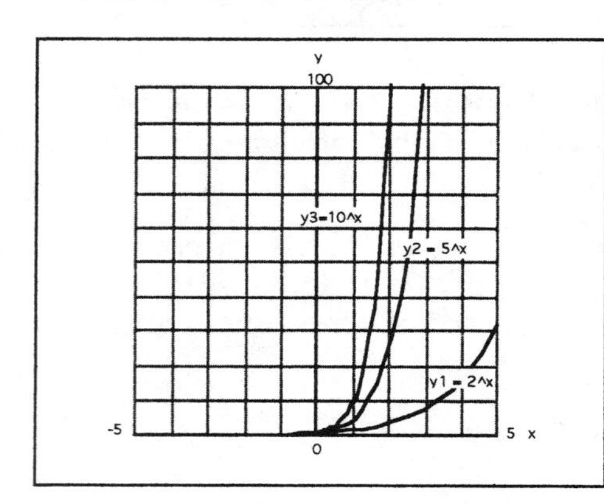

13. Let $y_1 = 2^x$, $y_2 = 5^x$ and $y_3 = 10^x$

a. $C = 1$ for each case; $a = 2$ for y_1, $a = 5$ for y_2 and $a = 10$ for y_3.

b. Each represents growth; y_1's value doubles, y_2's value is multiplied by 5 and y_3's value is multiplied by 10.

c. All three graphs intersect at (0,1)

d. In the first quadrant the graph of y_3 is on top, the graph of y_2 is in the middle and the graph of y_1 is on the bottom.

e. All have the graph of $y = 0$ (or the x-axis) as their horizontal asymptote.

f. Small table for each:

x	y_1	y_2	y_3
0	1	1	1
1	2	5	10
2	4	25	100

g. The graphs of the three functions are given in the diagram above on the left. They indeed confirm the answers given to the questions asked in parts **a.** through **f.**

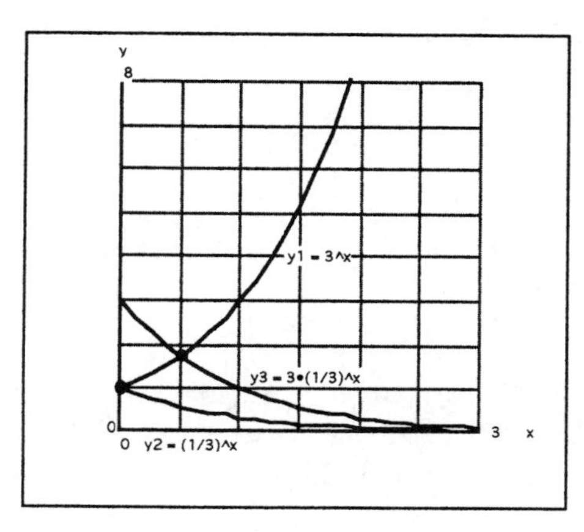

15. Let $y_1 = 3^x$, $y_2 = (1/3)^x$ and $y_3 = 3\bullet(1/3)^x$

a. C = 1 for y_1 and y_2 and C = 3 for y^3.

b. y_1 represents growth; y_2 and y_3 represent decay.

c. y_1 and y_3 intersect at (0.5, 30.5); y_1 and y_2 intersect at (0, 1).

d. The graph of y_1 is on top; the graph of y_3 is in the middle and the graph of y_2 is on the bottom.

e. All the graphs have the x-axis as their horizontal asymptote.

f. A small table for each:

x	y_1	y_2	y_3
0	1	1	3
1	3	1/3	1
2	9	1/9	1/3

g. The graphs of these three functions are given above on the left and they confirm the answers given to the questions in parts **a.** through **f.**

17. The function P goes with graph C; the function Q goes with graph A; the function R goes with graph B and the function S goes with graph D.

Exercises for Section 5.4

19. a. $Q(T) = 1000\bullet3^T$ **b.** $Q(T) = 1000\bullet1.3^T$
b. $Q(T) = 1000\bullet1.03^T$ **d.** $Q(T) = 1000\bullet1.003^T$

21. a. $P(t) = 150\bullet3^t$ **b.** $P(t) = 150 - 12t$ **c.** $P(t) = 150\bullet0.93^t$ **d.** $P(t) = 150 + t$

23. In 2020 it would be 1.4•250 = 350 million; in 2050 it would be 1.4•350 = 490 million. The general formula is P(t) = 250•1.4t, where t represents the number of 30 year periods from 1990. Graphing this function along with that of y = 1000 helps one to see that the point of intersection is approximately at t = 4.12•30 = 123.6 years.

25. a. $A(n) = A_o\bullet0.75n$, where n measures the number of years from the original dumping of the pollutant and A_o represents the original amount of pollutant.

b. Graphing y = 0.01 and y = 0.75n and computing their intersection gives n ≈ 16.008 years. (Note that the value of A_o is not directly relevant.)

27. a. $1.007^{12} = 1.087$ and thus the inflation rate is 8.7% per year.

b. $(1+r)^{12} = 1.05$ means that $r = (1.05)^{1/12} - 1 \approx 1.0041$. Thus the rate is about 0.4% per month.

Exercises for Section 5.5

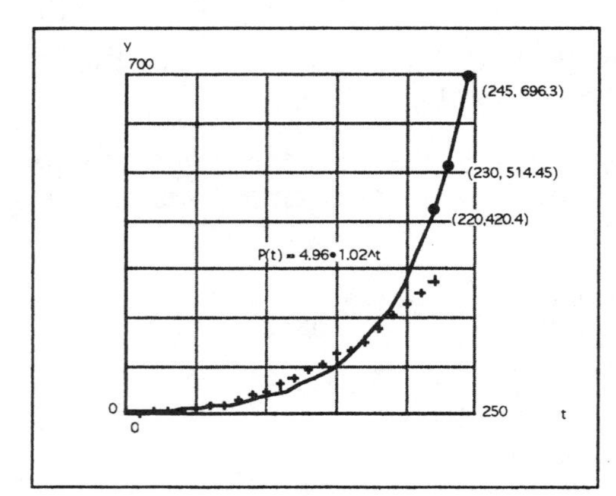

29. a. The best exponential fit of this data is given by:

$$P(t) = 4.96 \bullet 1.02^t$$

where P(t) measures the population of the US in millions t years since 1780 (following the scale given in the text).

The annual growth factor is 1.02 and the annual growth rate is 2% and the estimated initial population in 1790 is 5.06 million (approximately 1.3 times the actual value).

b. The graph of this function and that of the data is given in the diagram above on the left. For the first 50 years, the estimated population is only slightly larger than the actual one. Then for the next 100 years the real population is larger. Thereafter, the estimated population grows much more quickly than the actual one. For example, it predicts 420.4 million in 2000.

c. In 2010 the model predicts 515.5 million and in 2025 it predicts 696.3 million.

31. a. In the figure below on the left, the data is plotted along with the best-fitting straight line and exponential curves. It is clear from inspection of the graphs and the data that the exponential graph fits the data much better than the linear graph.

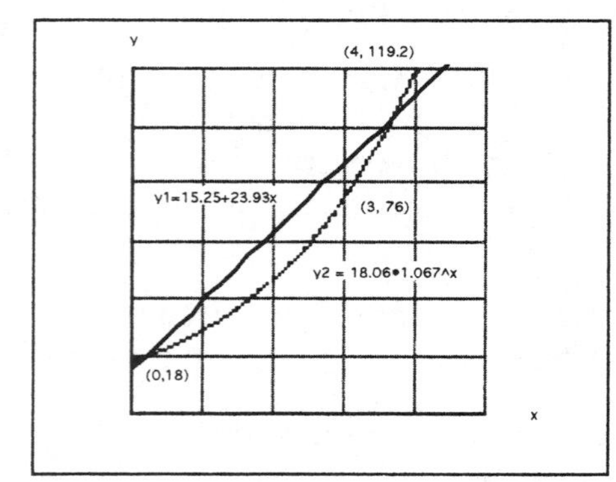

b. Using the data only from 1995 and 1999 to get the best-fit exponential curve gives:

$$U_1(x) = 18 \bullet 1.604^x$$

as the best-fitting exponential. Here x measures years since 1995 and $U_1(x)$ measures the number of users in millions in year x. Using this formula, the annual growth factor is 1.604 and the annual growth rate is 60.4%.

c. Using all the data, the best-fitting exponential formula (rounding numbers to two decimal places) is: $U_2(x) = 18.1 \bullet 1.61x$, where x = the number of years since 1995 and $U_2(x)$ = the number of users in millions in year x. With this formula, the annual growth factor is 1.61 and the annual growth rate is 61%. These are slightly higher than what one gets by using just the first and last data points as in part **b.**

d. In 2010, x = 15 and thus the formula obtained in part **c.** predicts that there will be $U_2(15) \approx$ 22,912 million internet users. In 2025, x = 30 and thus it is predicted that there will be $U_2(3) \approx$ 29,003,797 million users.

e. Using the exponential model derived in **#29** for US population, one could say that the US population in 1995 was approximately 350 million and that 76 million is 21.7% of that population. The same model says that in 1999 the US population was approximately 379 million and 119.2 million is approximately 31.5% of that population

[COMMENT: Using the actual figures for 1995 and 1999 would give respectively 29% and 43.8% for the amount of internet users.]

f. Using the figures given in **#29. c.** and in **#31. d.**, we see that in both 2010 and 2025 the predicted number of internet users far exceeds the predicted population of the US. The two models, therefore, are not very compatible that far into the future.

33. a. Below are the tables with the salaries for each year rounded to the nearest penny, where A(t) represents the salary that would be given by Aerospace Engineering Corp. in year t and where B(t) represents the salary that would be given by Bennington Corp. in year t.

t	A(t)	B(t)	t	A(t)	B(t)
0	50,000	35,000.00	5	60,000	56,367.85
1	52,000	38,500.00	6	62,000	62,004.64
2	54,000	42,350.00	7	64,000	68,205.10
3	56,000	46585.00	8	66,000	75,025.61
4	58,000	51243.50			

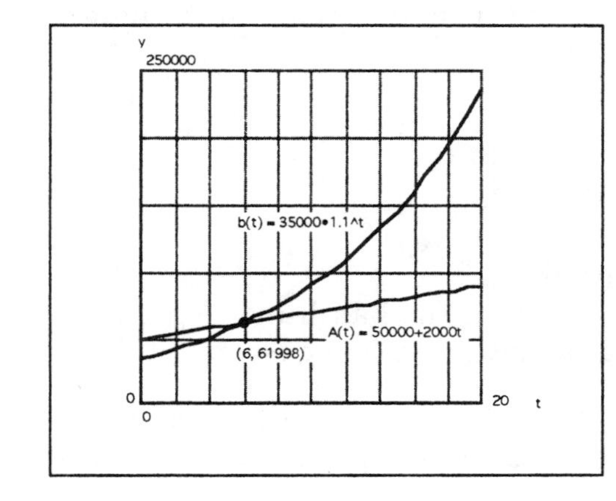

b. The formulae would be:

$$A(t) = 50,000 + 2000t$$

and

$$B(t) = 35,000 \bullet 1.1^t$$

where t measures years since hiring and A(t) and B(t) represent salaries in dollars after t years of employment.

c. The graphs of these functions are given in the diagram to the left. The location and coordinates of the intersection point are given.

d. The salary from Bennington would exceed that from Aerospace at the start of year 6. (The point of intersection occurs just before year 6.)

e. A(10) = \$70,000 and B(10) ≈ \$90,780.99; A(20) = \$90,000 and B(20) ≈ \$235,462.50

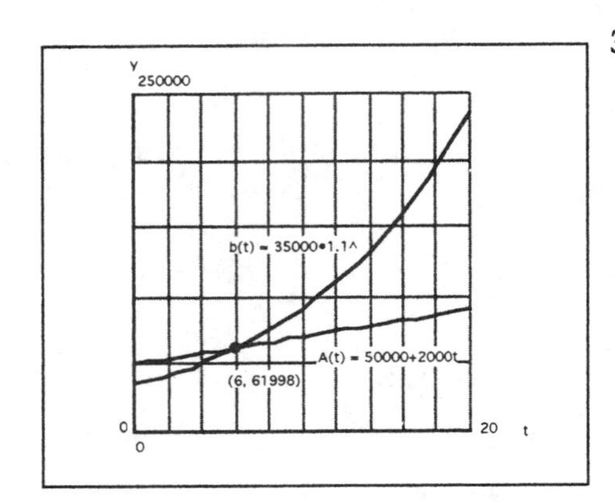

35. a. The linear model: LS(t) = 100 – 10t, where t is measured in minutes and LS(t) is measured in percent chance of survival. LS(t) ≤ 50 if 100 – 10t ≤ 50 or 5 ≤ t. Thus, using this model, one's chances are less than 50% after 5 minutes. The graph for this model is included in the figure on the left.

b. ES(t) = 100•0.90t, where ES(t) and t are respectively measured in the same units as LS(t) and t are in part **a**. ES(t) ≤ 50 if $100\bullet0.90^t \le 50$ or $0.90^t \le 0.5$ or t•log(0.90) ≤ log 5 or –0.4576•t ≤ -0.3010 or t ≥ 6.57 minutes. Thus, one's chances of survival are less than 50% in the exponential model after 6.57 minutes. The graph of this model is also included in the figure on the left..

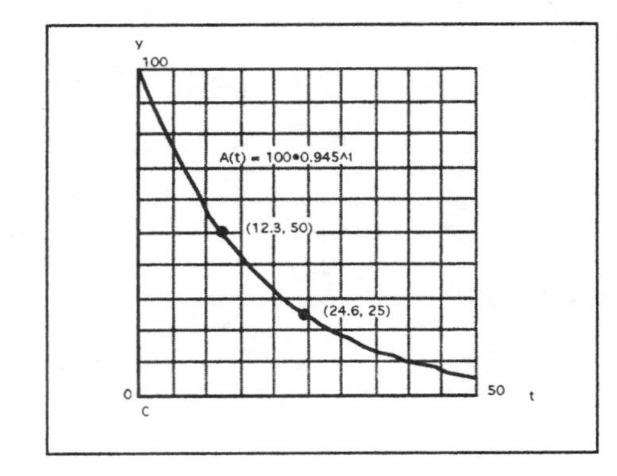

37. The graph goes through the points (0, 100) and (12.3, 50), where the first coordinate is measured in years since the tritium weights 100 kg. and the second is measured in kilograms.

Using the graphing utility gives:

$$A(t) = 100\bullet0.945^t$$

as the best-fitting exponential formula. The graph of this function is given in the figure on the left.

39. a. 1/32nd or 3.125% of the original dosage is left.

b. **i.** Approximately 2 hours.
ii. $A(t) = 100\bullet0.5^{t/2}$
iii. 5•2 = 10 hours and 100/32 = 3.125 milligrams
iv. There are many answers. You should mention the half-life and present a graph to make the drug's behavior clear to any prospective buyer.

41. a. R = 70/5730 ≈ 0.012 percent **b.** R = 70/11460 ≈ 0.0061 percent
c. R = 70/(5/31,536,000) = 441,504,000 percent **d.** R ≈ 220,752,000 percent

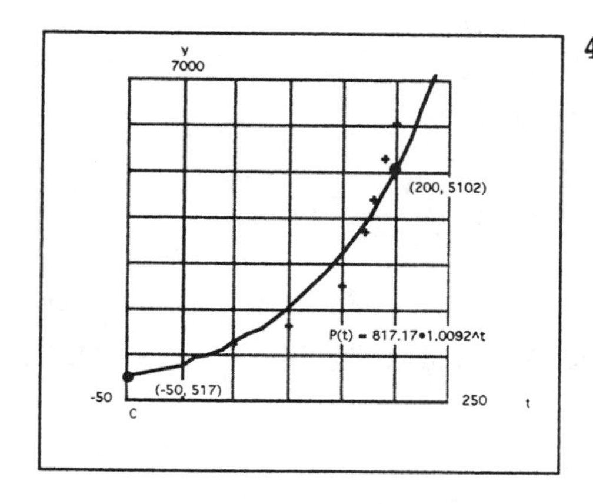

43. b. $P(t) = 817.17 \bullet 1.0092t$ is the best-fitting exponential function formula (after rounding off to 2 decimal places) where t is measured in years since 1800 and P(t) is measured in millions. Its graph along with that of the data table are given in the diagram to the left.

c. The 817.17 is the model's value for the world's population size in millions in 1800 (when $t = 0$). (Note that it is approximately 163 million smaller than the actual size.) The domain of the function P, technically is any real number but, concretely, its values should not go much farther back than 1800 or much past 1980 since (as can be seen from inspecting the graph given above) the actual growth is faster after 1980 than the model predicts. The range in that domain is from 817.17 million to 4248.31 million. (If one carries the domain to 2000, then the range goes up to 5102.24 million in the model.)

d. The base, 1.0092, means that the growth rate is 0.92%. This rate is measured in millions of persons per year.

t	P(t)
-50	516.95
120	2452.34
225	6414.93
250	8065.34

e. i. Using a calculator gives the P(t) values in the table to the left. Your eyeball estimates should be close to these.

ii. Using a calculator or the graph of the model one can see that 1 billion was reached during 1823 (or more precisely, 22.05 years after 1800); 3.2 billion was reached during early 1950 (or more precisely, 149.07 years from 1800); 4 billion was reached almost half way through 1974 (or more precisely, 173.42 years from 1800); and 8 billion will be reached early during 2050 (or more precisely 249.11 years from 1800).

iii. As can be seen from the answers in **ii.** it takes approximately 75.69 years for the population to double.

45. a. 10 half lives gives a $0.5^{10} = 1/1024$ of the original or a $1023/1024 \approx 99.9\%$ reduction.

b. $A(2) = 0.25A^{o}$, $A(3) = 0.125A_o$, $A(4) = 0.0625A_o$. After n half lives, the amount left is $A(n) = 0.5^n \bullet A_o$.

47. a. Let M(t) = the size of the tumor in grams after t months. Then M(3) =20, M(6) = 40, M(9) = 80 and M(12) = 60, all in grams. In general $M(t) = 10 \bullet \left(2^{1/3}\right)^t = 10 \bullet 1.2599^t$.

b. Solving M(t) = 2000 for t gives t ≈ 22.93 month or nearly 2 years.

c. 2000/10 = 200 and thus at 2000 grams, it is 20,000% of its original size.

49. a. i. $A(t) = 5000\bullet(1.035)^t$ **ii.** $B(t) = 5000\bullet(1.0675)^t$ **iii.** $C(t) = 5000\bullet(1.125)^t$.
b. A(40) = 19,796.30 B(40) = 68,184.45 C(40) = 555,995.02

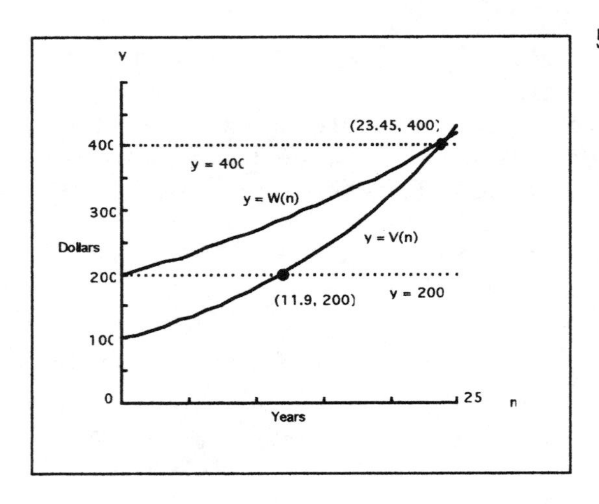

51. a. $V(n) = 100\bullet1.06^n$. Solving V(n) = 200 for n graphically gives n ≈ 11.90 years.

b. $W(n) = 200\bullet1.03^n$. Solving W(n)=400 for n graphically gives n ≈ 23.45 years.

c. The graphs are y = V(n) and y = W(n) are in the diagram to the left. It looks like the two intersect each other at approximately n = 24 years and that V's values are larger thereafter. Using algebra, one gets more precisely that the intersection point is (24.13,408.28).

53. Since 1.00/1.06 = 0.943, then the value of a dollar after t years of such inflation is given by $V(t) = 0.943^t$. Solving V(t) = 0.50 for t gives t ≈ 11.8 years.

then P = $343.06.

55. a. and **b.** The data plot shows that it represents a growth phenomenon and suggests an exponential form with the general pattern $D = Ca^t$, where D is measured in billions of dollars and t is measured in 5 year periods from 1970. The formula for the best-fitting exponential curve (rounded off) is $D = 334.71\bullet1.73^t$. Both the data and the exponential curve are plotted in the diagram below on the left.

Graph of the best-fit exponential model and of the data

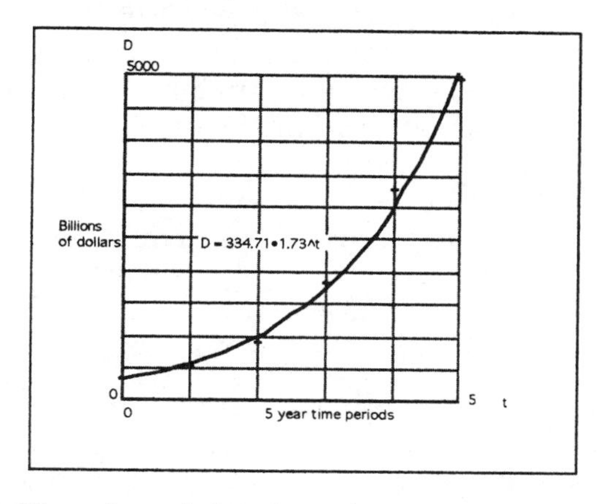

c. Table for time periods and debt ratio

Period	Debt Ratio
70 to 75	1.499
75 to 80	1.680
80 to 85	2.010
85 to 90	1.788
90 t 95	1.519

The debt ratios are not that constant. The average of these ratios is 1.699. (The ratios and their average are all rounded off to 3 decimal places.) Thus, G = 1.699 is the average growth factor.

d. The plot of the data along with the exponential function that uses 361 as the starting amount and G = 1.699 as the base is given below on the left.

e. If $4961 = 361 \bullet G^5$, then $G \approx 1.689$ (when rounded to three decimal places). The plot of the data along with the exponential function that uses 361 as the starting amount is given below on the right.

Plot for # **55. d.**

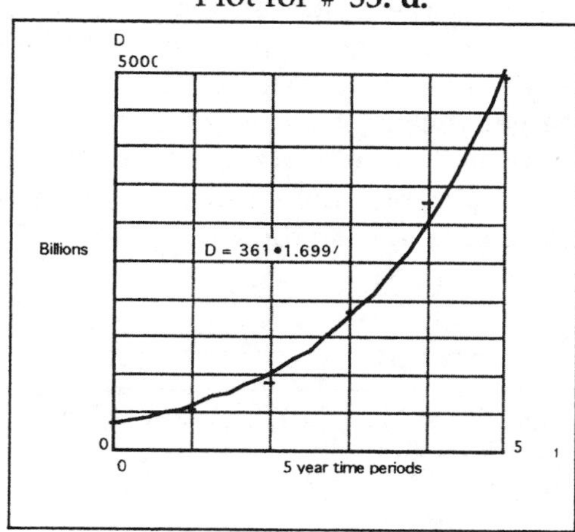

Plot for #**55. e.**

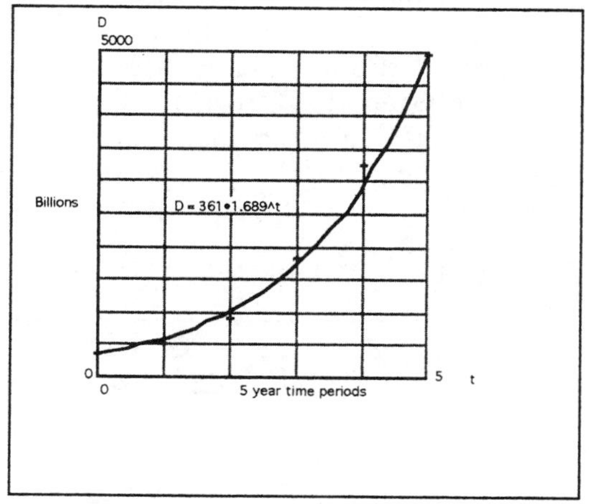

f. Your answer as to which is the best-fitting model could be any of these. The original best- fit model along with the other two functions predict the following values (rounded off to two decimal places) for 2000 and 2005.

Model	2000	2005
Best-fit model	8,973.16	15,523.56
Model from part **c.**	8,682.95	14,752.34
Model from part **e.**	8,380.79	14,155.16

Comments about these functions and their graphs: The function computed in **#55.c.** uses a smoothing out of the growth ratios. The one computed in **#55.e.** goes through the initial and final data points. Note that the graph of the best-fit exponential function starts out lower than the other two, but, since it has a faster growing rate than either of them, its y values will eventually surpass the y values of the other two. The graph of the function computed in **#55.c.** is the top graph at the start and stays there until the best-fit graph surpasses it around t = 4.27. The graph of the function computed in **#55. e.**, except at t = 0, is always the below the graph asked for in **#55.c.** and the best-fit function surpasses it around t = 3.21. (These values are obtained by blowing up the graphs at various spots between t = 0 and t = 5.) It is impossible to get such information from overlaying the three graphs between t = 0 and t = 5. They are too close together in that range.

57. a. $M_{new}(n) = 10^n$, where n measures the number of rounds and $M_{new}(n)$ measures the number of people participating in the n rounds of recruiting. Note that this formula assumes that all who are recruited stay and that all recruits are distinct.

b. $M_{Total}(n) = 1 + 10 + \cdots 10^n$.

c. $M_{Total}(5) = 11{,}111$ but only 11,110 of those stem from the originator. After 10 rounds the number recruited (not including the originator) would be 11,111,111,110.

d. Comments will vary but all will probably note how fast the number of recruits needed grows and how the amounts expected are not quite what one would have thought from the advertisements. For example, it does not seem intended that, in order for a person to get the discount, each person in the group that the person has recruited will in turn have to recruit a new set of 10 persons, *etc*. Otherwise, the whole country, if not the whole world, would have to join in order that anyone would be able to get a discount.

59. a. Here is the table asked for:

	Females		Males	
t	female births	total females	male births	total males
0		100		100
1	200	300	200	300
2	600	900	600	900
3	1800	2700	1800	2700
4	5400	8100	5400	8100
5	16200	24300	16200	24300
6	48600	72900	48600	72900
7	145800	218700	145800	218700
8	437400	656100	437400	656100

b. $F(t) = 100 \bullet 3^t$ and $M(t) = 100 \bullet 3^t$, where t is measured in 6-month periods and the function values are female and male cat counts respectively.

c. Since t counts 6-month periods, the number of cats after 5 years is when t = 10. Now F(10) = M(10) = 5,904,900 and thus there would be 11,809,800 cats altogether. The increase is one of 5 orders of magnitude over 5 years.

d. Under these conditions, in 5 years there would be 25•310 = 1,476,225 males and the same number of females and thus there would be 2,952,450 cats in all.

61. a. Below is the graph of the white blood cell counts on a semi-log plot. The data from October 17th to October 30th seem to be exponential, since the data in that range seem to fall along a straight line in the plot. The progress from September 30th to October 5th could be considered exponential decay but the time period is rather short.)

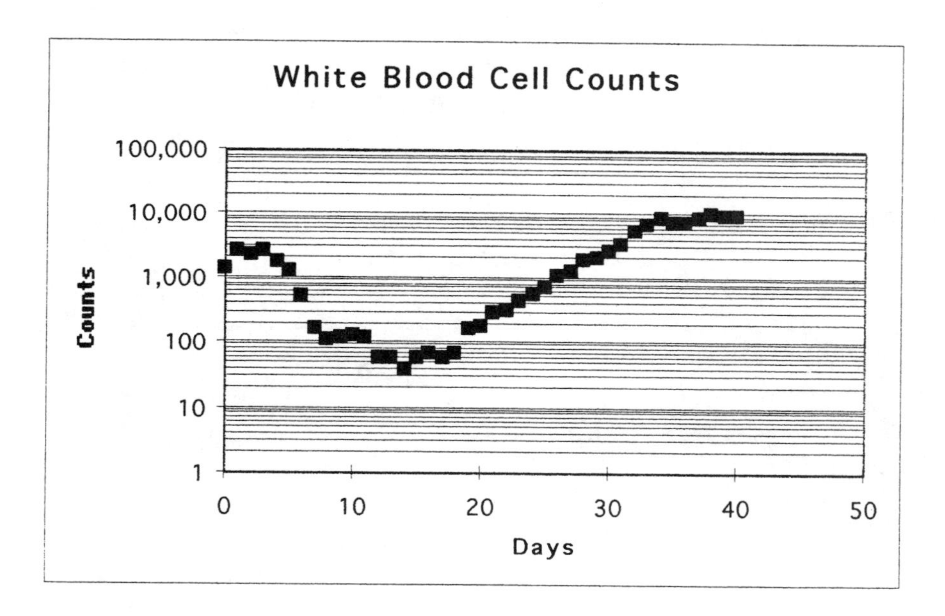

b. On the next page are two graphs of the E-Coli counts, one has regular horizontal and vertical axes and the other is a semi-log plot. This data looks very exponential from the third to the thirteenth time periods in the regular plot and fairly exponential in the semi-log plot (since that plot looks rather linear). For contrast, the regular linear plot is given as well.

Semi log plot for **#61.b**

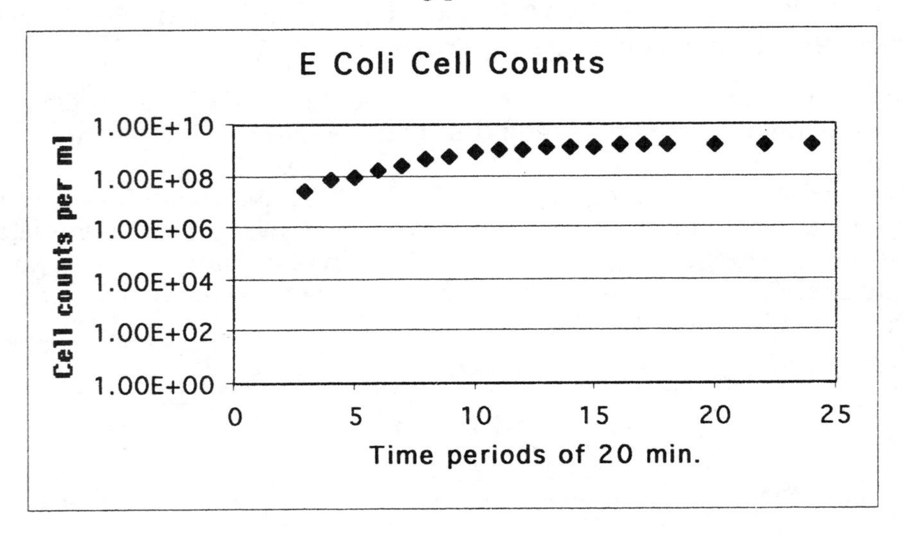

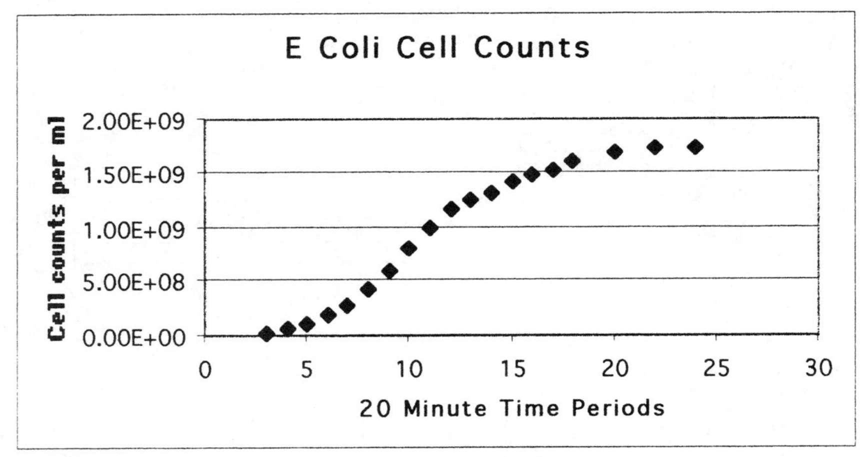

Regular plot for **#61 b.**

63. Note that all numbers in the answers below are eyeball estimates.

a. This is a semi-log plot.

b. In 1929 it was 400; in 1933 it was 40; this latter was in the depression era.

c. It was first 1000 in 1966; it was 3000 first in 1994; it was 6000 first in 1997.

d. A straight line graph on a semi-log coordinate system means that the growth is exponential, since the equation of that straight line is $\log(y) = mx + b$ and this means that $y = 10^{mx+b}$ and the latter can be written in the form $y = Ca^x$, with $C = 10^b$ and $a = 10^m$. (Note that in a semi-log plot even though the vertical distances are log(y) distances, the vertical scale is labeled as if they were the actual y distances.)

e. In 1980 the DJA was 900 and in 1997 it was 7000. Thus, if x is measured in years since 1980, we have that $7000 = 900 \bullet a^{17}$ and thus $a \approx 1.128$. This would mean that the annual rate of growth over that period was 12.8%.

f. Similarly, solving $7000 = 40 \bullet a^{64}$ gives $a = 1.084$. Thus the annual growth rate over that time period was 8.4%.

Exercises for Section 6.1

1. Your estimates may be different for each interest rate. The ones given here are based on rough approximations of the position of 300 between the two relevant data entries.

a. approximately 37 yrs. **b.** approximately 22 yrs. **c.** approximately 15 yrs.

3. Eyeball estimates will vary. One set of guesses is: **a.** 1.3 hrs. **b.** 2.4 hrs. **c.** 4.3 hrs.

5. a. 0.001 **b.** 10^6 **c.** $10^{1/3}$ **d.** 1 **e.** 10 **f.** 0.1

7. If $w = \log(A)$ and $z = \log(B)$, then $10^w = A$ and $10^z = B$.
Thus $A/B = 10^w/10^z = 10^{w-z}$ and
therefore $\log(A/B) = \log(10^{w-z}) = w - z = \log(A) - \log(B)$, as desired.

9. a. $\log\left(\frac{K^3}{(K+3)^2}\right)$ **b.** $\log\left(\frac{(3+n)^5}{m}\right)$

11. Let $w = \log(A)$. Then $10^w = A$ and thus $A^p = (10^w)^p = 10^{wp}$ and then
$\log(A^p) = w \bullet p = p \bullet w = p \bullet \log(A)$, as desired.

13. a. 100 **b.** 999 **c.** $10^{5/3} \approx 46.416$ **d.** 0.5 **e.** 1/9 **f.** none

15. a. $\log(3)/\log(1.03) \approx 37.17$ years
b. $\log(3)/\log(1.05) \approx 22.51$ years
c. $\log(3)/\log(1.07) \approx 16.24$ years

17. a. $t = \ln(2)/\ln(5) = \log(2)/\log(1.5) \approx 1.71$ 20-minute time periods or 34.19 mins.
b. $t = \ln(10)/\ln(1.5) = \log(10)/\log(1.5) \approx 5.68$ 20-minute time periods or 113.58 mins.

19. a. $B(t) = B_o(0.5)^{0.05t}$ – where t is measured in minutes and B(t) and B_o are measured in some weight unit. None is specified in the problem.

b. In one hour, $t = 60$ and thus $B(60) = B_o(0.5)^{0.05 \bullet 60} = 0.125 \bullet B_o$ or 12.5% is left.

c. It its half-life is 20 minutes then its quarter life is 40 minutes.

d. Solving 0.10 = 0.50.05t for t, we get $0.05t \bullet \log(0.10) = \log(0.5)$ or
$t = \log(0.1)/[\log(0.5) \bullet 0.05] \approx 66.439$ minutes.

21. a. $S(W) = 300 \bullet 0.9^W$ if $0 \le W \le 10$, where W is measured in weeks and S(W) is measured in dollars.

b. Solving $S(W) = 150$ for W means solving $0.5 = 0.9^W$ for W and thus $W = \log(0.5)/\log(0.9) \approx 6.58$ weeks. The selling price when it would be given to charity is $S(10) = 300 \bullet 0.9^{10} \approx \$104.60$

Exercises for Sections 6.2 and 6.3

23. In general if the nominal rate is 8.5%, then $A_k(n) = 10{,}000\bullet(1 + 0.085/k)^{kn}$ gives the value of \$10,000 after n years if the interest is compounded k times per year and $A_c(n) = 10{,}000\bullet e^{0.085n}$ gives that value if the interest is compounded continuously.

a. annually: $(1 + 0.085/1)^1 = 1.0850$ and thus the effective rate is 8.5%
b. semi-annually $(1 + 0.085/2)^2 \approx 1.0868$ and thus the effective rate is 8.68%
c. quarterly $(1 + 0.085/4)^4 \approx 1/0877$ and thus the effective rate is 8.77%
d. continuously $e^{0.085n} \approx 1.0887$ and thus the effective rate here is 8.87%

25. a. $U(x) = 10\bullet(1/2)^{x/5}$, where x is measured in billions of years.

b. $x = 5\bullet\log(0.1)/\log(0.5) \approx 16.6096$ billion years.

27. a. $t = \log(2)/\log(1.12) \approx 6.12$ years

b. $t = \log(2)/\log((1+0.12/4)^4) \approx 5.86$ years.

c. $t = \ln(2)/0.12 \approx 5.78$ years

29. a. Using the method of the text in this section we find that $e^{0.0343} = 1.0349$ and thus $r \approx 0.0349$ or 3.49%

b. $e^{0.046} \approx 1.0471$ and thus $r \approx 0.047$ or 4.71%

31. a. $10^n = 35$ **b.** $N = N_o e^{-kt}$

33. Let $w = \ln(A)$ and $z = \ln(B)$. Thus $e^w = A$ and $e^z = B$. Therefore $A\bullet B = e^{w+z}$ and therefore $\ln(A\bullet B) = w + z + \ln(A) + \ln(B)$, as desired.

35. a. $\ln\left(\sqrt[4]{(x+1)(x+3)}\right)$ **b.** $\ln\left(\frac{R^3}{\sqrt{P}}\right)$ **c.** $\ln\left(\frac{N}{N_o^{\,2}}\right)$

37. a. $x = \ln(10) \approx 2.303$ **b.** $x = \log(3) \approx 0.477$ **c.** $x = \log(5)/\log(4) \approx 1.161$
d. $x = e^5 \approx 148.413$ **e.** $x = e^3 - 1 \approx 19.086$ **f.** no solution (ln increases)

Exercises for Section 6.4

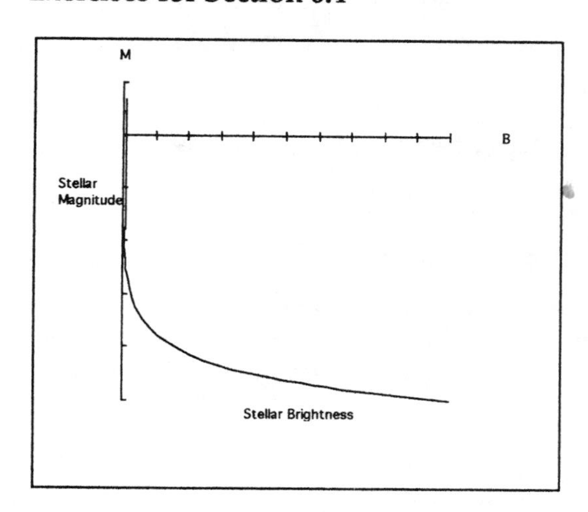

39. a. At $B = B_o$.

b. A rough sketch of its graph is given in the diagram on the left. (The scale is very spread out on it.) Its domain is all $B > 0$.

c. As the brightness increases the magnitude decreases. Thus a 6th magnitude star is less bright than a 1st magnitude star.

d. If the brightness goes from B to 5B then the magnitude M becomes $2.5 \bullet \log(5) \approx 1.75$ units less is size.

41. $dB = 10 \bullet \log(I/I_o) = 28$ implies that $I/I_o = 10^{2.8}$ and thus $I = 10^{-13.2}$ watts/cm^2 and 92 = dB has $I/I_o = 10^{9.2}$ and thus $I = 10^{-6.8}$ watts/cm^2. (Note that these answers assume, of course, that $I/I_o = 10^{-16}$ watts/cm^2.)

43. If the combined intensity is called Cm and the individual intensity is called In, then $Cm = 5 \bullet In$. Thus

$$dB(Cm) = 10 \bullet \log(5 \bullet In/I_o) = 10 \bullet \log(5) + 10 \bullet \log(I/I_o) \approx 6.99 + dB(In).$$

Thus, a set of quintuplet babies crying is about 7 decibels louder than one baby crying.

Exercises for Section 6.5

45. a. $N = 10 \bullet e^{0.044017t}$ b. $Q = 5 \bullet 10^{-7} \bullet e^{-2.631089A}$

47. a. decay   **b.** decay   **c.** growth

49. If $200 = 760 \bullet e^{-0.128h}$, then $h = \ln(200/760)/(-0.128) \approx 10.430$ km

51. The half-life is $\ln(1/2)/(-r) = \ln(2)/r = 100 \bullet \ln(2)/R \approx 69.3137/R$ which is approximately $70/R$

53. Your answer may be different. I will look for a curve that goes roughly through the middle of each cluster. Without a data table of values it is not possible to give the best-fit exponential with any precision.

55. Given: $n = n_o \bullet e^{\ln(2)\bullet t/T}$ and $11 = n_o \bullet e^{\ln(2)\bullet 2/T}$ and $30 = n_o \bullet e^{\ln(2)\bullet 22/T}$.

Therefore: $\ln(11) = \ln(n_o) + 2\bullet\ln(2)/T$ and $\ln(30) = \ln(n_o) + 22\bullet\ln(2)/T$

Thus $\ln(30) - \ln(11) = \ln(2)[22 - 2]/T$

or $T = 20\bullet\ln(2)/[\ln(30) - \ln(11)]\ n_o \bullet e^{\ln(2)\bullet t/T} \approx 13.817$ seconds

and $11 = n_o \bullet e^{[(\ln(30) - \ln(11))/10]}$ or $11 \approx n_o \bullet 1.1055$

Thus $n_o \approx 9.95$ or 10 neutrons in the needed round off.

57. a. One would have to divide each of the dollar amounts in the table by the number of people in the US in that year. The table furnished in Chapter 2 lists populations in millions for 1960, 1970, 1980, 1990 and 1995. We can interpolate for 1965, 1975 and 1985.

Doing these computations, one gets the following cost table (where costs per person are rounded off to the nearest dollar):

Year	1965	1970	1975	1980	1985	1990	1995
Cost in $ per person	214	358	606	1088	1799	2805	3770

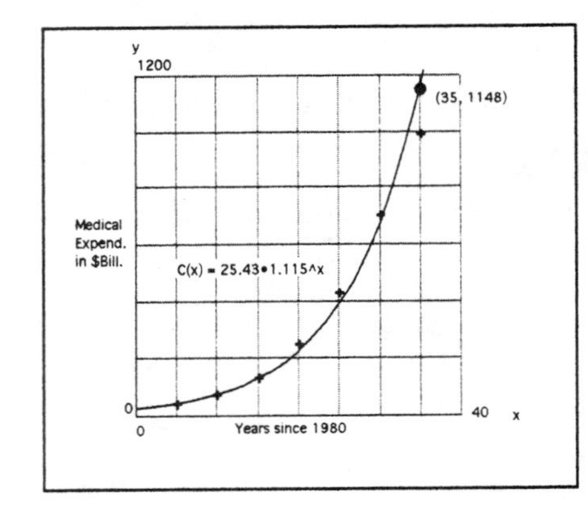

b. and **c.** From the scatter diagram of the total expenditure amounts over the indicated time period, it is fair to say that the growth is exponential. Using the graphing software gives that the best exponential curve fit has the equation:

$$C(x) = 25.43\bullet 1.115^x$$

where x measures years since 1960 and C(x) measures the total medicine expenditures in year x in billions of dollars. The model predicts that in 1995 the total expenditure will be approximately $1148 billion. The actual costs were $156 billion less.

Exercises for Section 6.6

59. a. $F(G(1)) = F(0) = 1$

b. $G(F(-2)) = G(-3) = 4$

c. $F(G(2)) = f(0.25) = 1.5$

d. $F(F(0)) = F(1) = 3$

e. $F\circ G(x) = F(G(x)) = F\left(\frac{x-1}{x+2}\right) = 2\left(\frac{x-1}{x+2}\right) + 1 = \left(\frac{3x}{x+2}\right)$

f. $G\circ F(x) = G(2x+1) = \frac{(2x+1)-1}{(2x+1)+2} = \frac{2x}{2x+3}$

61. a. $f(g(1)) = f(0) = 2$
b. $g(f(1)) = g(1) = 0$
c. $f(g(0)) = f(1) = 1$
d. $g(f(0)) = g(2) = 3$
e. $f(f(2)) = f(3) = 0$

63. a. g(f(2)) = g(0) = 1 **b.** f(g(-1)) = f(2) = 0
c. g(f(0) = g(4) = -3 **d.** g(f(1)) = g(3) = -2

65. a. V = (4/3)π(5+p)3 **b.** V = g(f(p))

67. a. T = 32 – 5s **b.** $S(x) = \left[1 - \frac{1}{2} \bullet \left(\frac{x}{20}\right)^2\right] S_d$

c. At the middle of the road, x = 0 and S(0) = S_d.
At the edge of the road, x = 20 and S(20) = 0.55S_d.

d. $T(x) = 32 - 5 \bullet \left[1 - \frac{1}{2} \bullet \left(\frac{x}{20}\right)^2\right] S_d$ e. T(0) = 32 – 2.5S_d.

69. a. $f(g(x)) = f\left(\frac{x+1}{2}\right) = 2\left(\frac{x+1}{2}\right) - 1 = x + 1 - 1 = x$

$g(f(x)) = g(2x - 1) = \frac{(2x-1)+1}{2} = \frac{2x}{2} = x$

b. $f(g(x)) = f\left(\frac{x^3 - 5}{4}\right) = \sqrt[3]{4\left(\frac{x^3-5}{4}\right) + 5} = \sqrt[3]{x^3} = x$ and

$g(f(x)) = g\left(\sqrt[3]{4x+5}\right) = \frac{\left(\sqrt[3]{4x+5}\right)^3 - 5}{4} = \sqrt[3]{x^3} = x$

71. They are inverse functions by definition, *i.e.*, $e^{\ln(x)} = x$ and $\ln(e^x) = x$. See the tables of properties for these two functions.

73. a.

x	$f^{-1}(x)$
5	-2
1	-1
2	0
4	1

b.

x	$f^{-1}(x)$
5	0
3	1
2	-2
-7	4

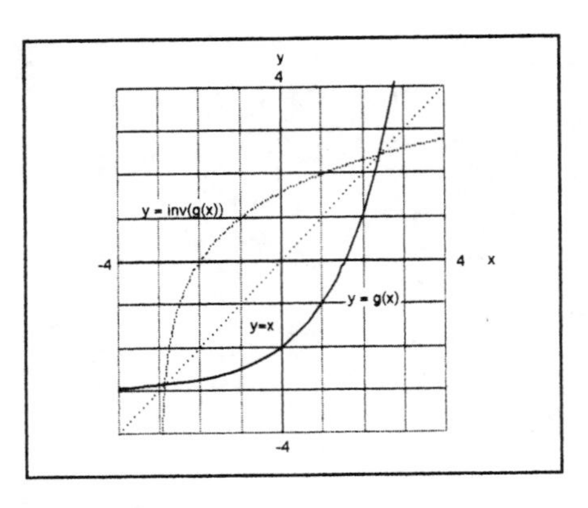

75. The graph of g and its inverse along with the graph of y = x (the line of symmetry) are all in the figure on the left.

77. a. $x = 4y - 7$ or $4y = x + 7$ or $y = (x+7)/4$
Thus $f^{-1}(x) = (x+7)/4$

b. $x = 8y^3 - 5$ or $x + 5 = y^3$ or $y = (x+5)^{1/3}$
Thus $g\text{-}1(x) = (x+5)^{1/3}$

c. $x = (y+1)/(2y)$ or $2xy = y + 1$ or $y = 1/(2x+1)$ Thus $h^{-1}(x)=1/(2x+1)$

79. a. The formula is $T = A + Ce^{-kt}$. The data are: A = 70 °F; at t = 0 T = 375 °F and at t = 30 T = 220 °F. Thus, when t = 0 we have 375 = 70 + C or C = 305. Also, at t = 30 we have $220 = 70 + 30\ e^{-30k}$ and therefore $150/305 = e^{-30k}$. Hence $k = \ln(150/305)/(-30) \approx 0.237$. Thus $T = 70 + 305e^{-0.237t}$ is the law of cooling here.

b., **c.** and **d.** $t = \frac{1}{0.237}\ln\left(\frac{305}{T-70}\right)$ is the inverse function; moreover at T = 370 we have t = 0.6974 and at T = 374 we have t = 0.0139 and at T = 375 we have t = 0.

e. and **f.** The graphs of T and its inverse function are given below on the left and right respectively.

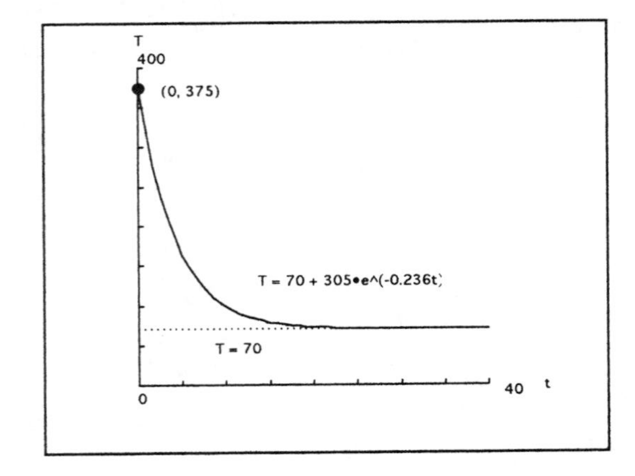

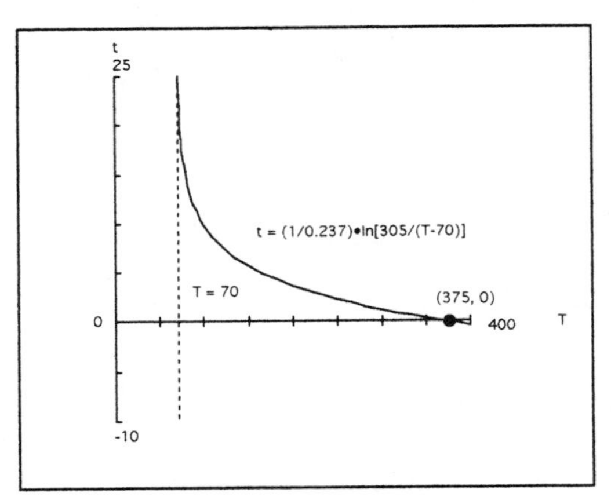

<><><><><><><><><><><><><><><><><><><><><><><><><><><><><><><><><><><>

Exercises for Section 7.1

1. a. $S = 1.26 \bullet 10^{-19}\ m^2$ **b.** $V = 4.19 \bullet 10^{-30}\ m^3$ **c.** $S/V = 3.0 \bullet 10^{10}\ m^{-1}$

3. a. S is multiplied by 16; V is multiplied by 64.
b. S is multiplied by n^2; V is multiplied by n^3.
c. S is divided by 9; V is divided by 27.
d. S is divided by n^2; V is divided by n^3.

5. a. The volume doubles if the height is doubled.
b. The volume is multiplied by 4 if the radius is doubled.

7.

radius	C	A	Circumference ratio	Area Ratio
R	$2\pi R$	πR^2	1 to 1	1 to 1
2R	$4\pi R$	$4\pi R^2$	2 to 1	4 to 1
3R	$6\pi R$	$9\pi R^2$	3 to 1	9 to 1
0.5R	πR	$0.25\pi R^2$	1 to 2	1 to 4

When you multiply the radius by k, you multiply the circumference by k.
When you multiply the radius by k, you multiply the area by k^2.

Exercises for Section 7.2

9. a. 20, 20 **b.** 40, -40 **c.** -20, -20 **d.** -40, 40

11. a. $Y = kX^3$ **b.** $k = 1.25$ **c.** Increased by a factor of 125

d. Divided by 8 **e.** $X = \sqrt[3]{\frac{Y}{k}}$

13. a. L is multiplied by 32 **b.** M is multiplied by 2^P

15. Given: $I(x) = k/x^2$ and $4 = I(6) = k/36$; thus $k = 144$ and $I(8) = 144/64 = 2.25$ watts per square meter and $I(100) = 144/10000 = 0.0144$ watts per square meter.

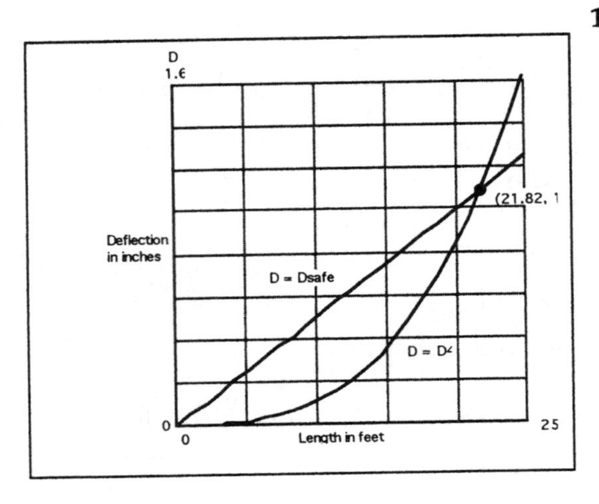

17. a. The graph of D_4 (with the deflections measured in inches and the plank length L measured in feet) is given in the diagram on the left. [Note that the domain used is between 0 and 25 ft instead of 20 ft. in order to demonstrate the safety issue asked about in part c. below]

b. The graph of D_{safe} is also given in the diagram on the left.

c. One notes that the safety deflections are well above the actual deflections for all values of L between 0 and 20. It would cease to be safe if the plank were longer than 21.82 ft. (This is the L value where D_{safe} and D_4 meet; it was solved for graphically).

19. In all the formulae below, k is the constant of proportionality.

a. $d = k \bullet t^2$ **b.** $E = k \bullet m \bullet c^2$ **c.** $A = k \bullet b \bullet h$ **d.** $R = k[O_2] \bullet [NO]^2$ **e.** $v = k \bullet r^2$

Exercises for Section 7.3

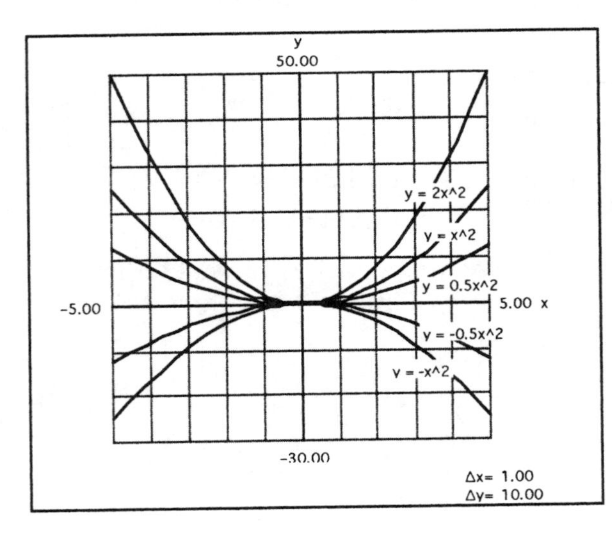

21. The graphs of the 5 functions are given in the diagram to the left. Each graph is labeled with its formula.

They are all basically the same shaped graph since they are graphs of functions of the form $y = ax^2$, for various values of a. The differences are in the effect of this constant multiplier. If $a > 0$ then the graph faces up If $a < 0$, then the graph faces down. The larger the absolute value of a, the narrower the opening of the graph.

Graphs of 1st four functions in #23

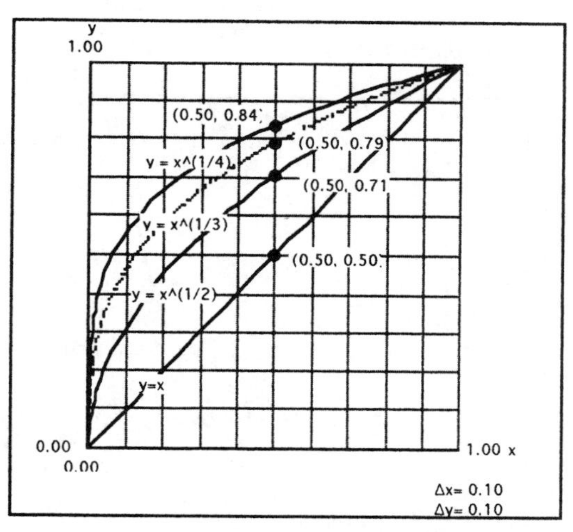

Graphs of 1st and last 3 functions in #23

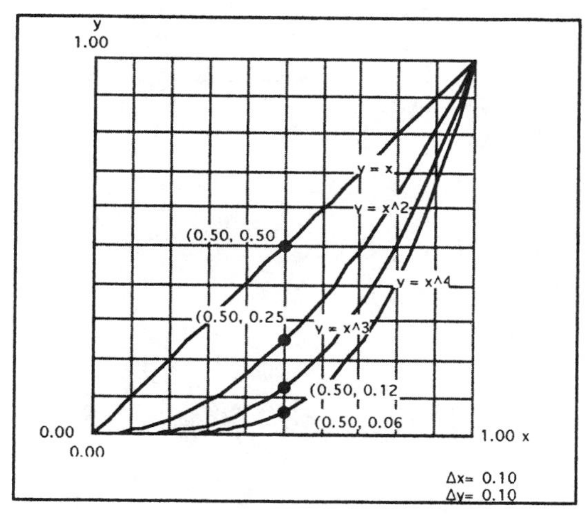

23. The graphs for the first four functions are given in the diagram above on the left. All go through (0,0) and (1,1). The points where $x = 0.5$ on each are marked with their y values for comparison. One notes that the smaller the fractional power the bigger the y value here.

The graphs of the first and the last three functions are given in the diagram above on the right. All graphs go through (0,0) and (1,1) and the points where $x = 0.5$ are marked off with their y values. The higher the power of x, the smaller the y value at $x = 0.5$

You may also notice that the fractional powers all have y values above or on the graph of $y = x$ and the whole number powers all have y values below or on the graph of the graph of $y = x$ over the interval [0, 1].

Exercises for Section 7.4

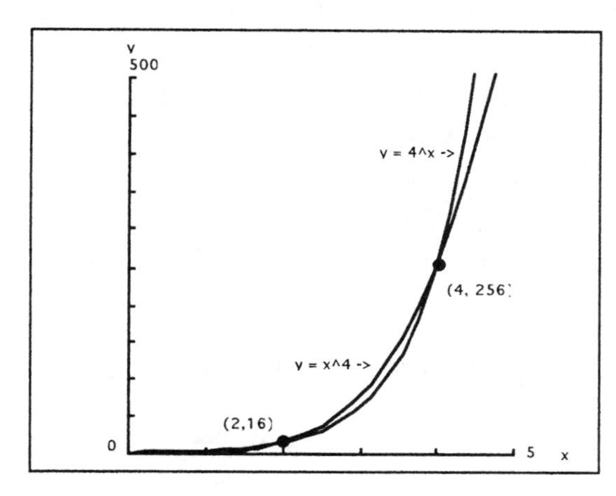

25. a. The two graphs intersect at (2, 16) and (4, 256). Thus, the functions are equal for $x = 2$ and $x = 4$. (See the figure on the left.)

b. As x increases both functions grow. For $0 \le x < 2$ we have that $x^4 < 4^x$; at $x = 2$ both have a y value of 16. From $2 < x < 4$, we have the $4^x < x^4$. For $x = 4$ they are again equal. For $x > 4$ we have that $x^4 < 4^x$. Lastly, the relative sizes of the y values stay that way as x keeps on increasing.

c. Eventually the graph of $y = 4^x$ dominates.

#27 Graphs of $y = 2^x$ and $y = x^2$

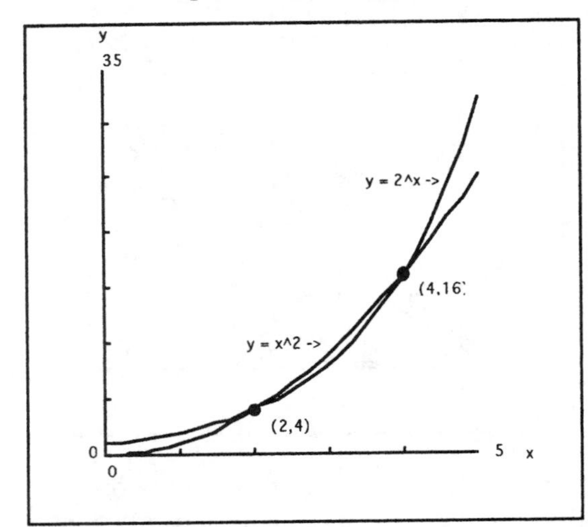

#27 Graphs of $y = 3^x$ and $y = x^3$.

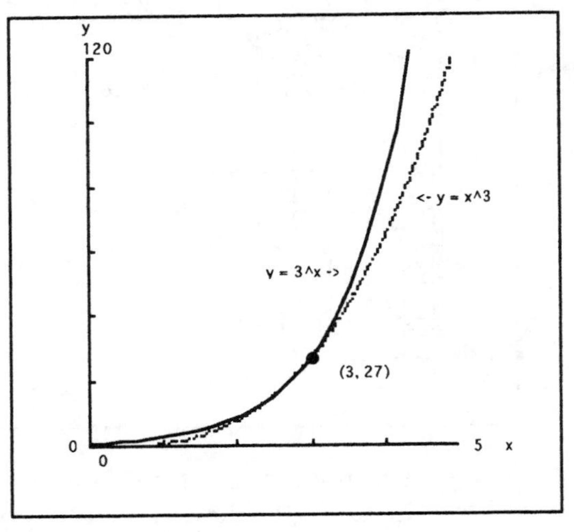

27. As can be seen from the figure above on the left, the graphs of $y = 2^x$ and $y = x^2$ intersect at $x = 2$, and $x = 4$. As can also be seen from the figure on the left, the graph of $y = 2^x$ is above the graph of $y = x^2$ $0 \le x < 2$ and for $x > 4$.

As can be seen from the figure above on the right, the graphs of $y = 3^x$ and $y = x^3$ intersect at $x = 3$. Moreover, the graph of $y = 3^x$ is above the graph of $y = x^3$ for all $x \ge 0$ except at $x = 3$.

29. $3 \cdot 2^x > 3 \cdot x^2$ if $0 < x < 2$ and if $x > 4$. Also, $3 \cdot 2^x < 3 \cdot x^2$ if $2 < x < 4$. (These are obtained from an inspection of their respective graphs and that these graphs have the same relative shapes as those in **#27** above except that all y values are now multiplied by 3.)

Exercises for Section 7.5

31. $I(d) = k/d^2$ and thus $I(4)/I(7) = 49/16 = 3.06$ -- it will be 3.06 times as great.

33. a. The volume becomes 1/3 of what it was. **b.** It becomes 1/n of what it was.
c. The volume is doubled. **d.** It becomes n times what it was.

35. a. Let L = wavelength of a wave, v = velocity of a wave, and t = time between waves, we have that $t = L/(k\sqrt{L}) = (1/k) \cdot \sqrt{L}$ and thus time is directly proportional to the square root of the wavelength of the wave.

b. If the frequency of the waves on the second day is twice that of the first then the waves will be 4 times as far apart as they were the previous day.

37. a. $x = k \cdot \frac{y}{z}$ **b.** Solving $4 = k \cdot \frac{16}{32}$ for k gives $k = 8$. **c.** $x = 8 \cdot 25/5 = 40$

Graph for # 39

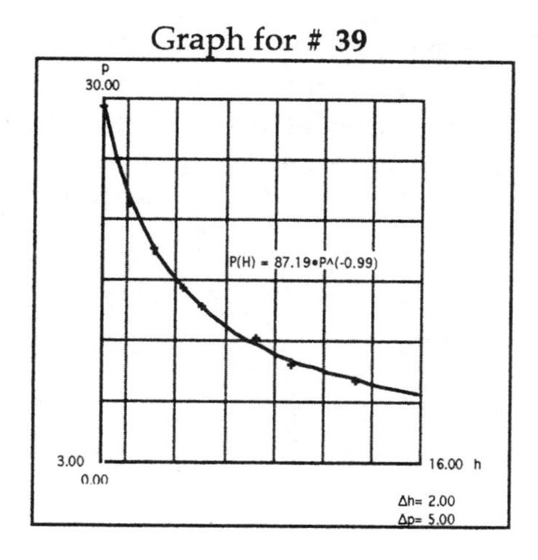

39. a. $H = k/P$ is suggested, where k is a proportionality constant.

b. The software gave $H = 87.19 \bullet P^{-0.99}$, which is very close.

Note that the fit is quite good as can be seen from the graph of the data and the best-fit graph as given above on the left.

Exercises for Section 7.6

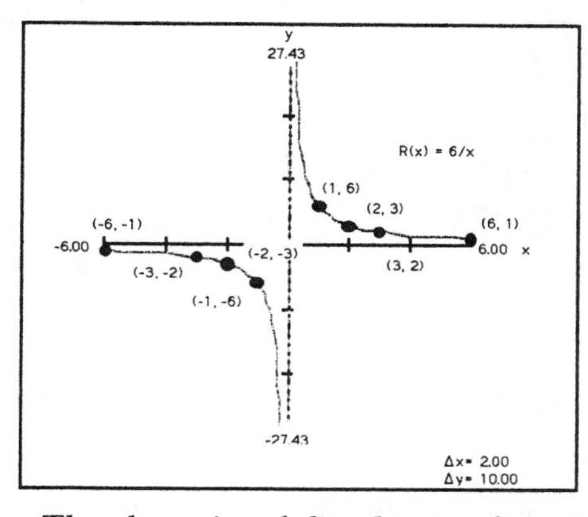

41. A table for r is given below.

x	r(x)
1	6
2	3
3	2
6	1
-1	-6
-2	-3
-3	-2
-6	-1

The domain of the abstract function is $x \neq 0$. For $x > 0$, as $x \rightarrow 0$ we have $r(x) \rightarrow +\infty$ and for $x < 0$, as $x \rightarrow 0$ we have $r(x) \rightarrow -\infty$.

43. a.

x	5x
-2	-10
2	10
-1	-5
1	5
0	0

x	x/5
-2	-0.4
2	0.4
-1	-0.2
1	0.2
0	0.0

x	1/x
-2	-1/2
2	1/2
-1	-1
1	1
0.5	2

x	5/x
-2	-5/2
2	5/2
-1	-5
1	5
0.5	10

b. The graphs of g and h are both straight lines through the origin with positive slope. In absolute value terms, the y values of g's graph are 25 times those of h. As for the graphs f and t: both have the x and y axes as asymptotes, *i.e.*, they approach but never touch these axes and both are confined to the first and third quadrants. The y values of the graph of f are 5 times the y values of the graph of t.

Graphs of $g(x)=5x$ and $h(x) = x/5$

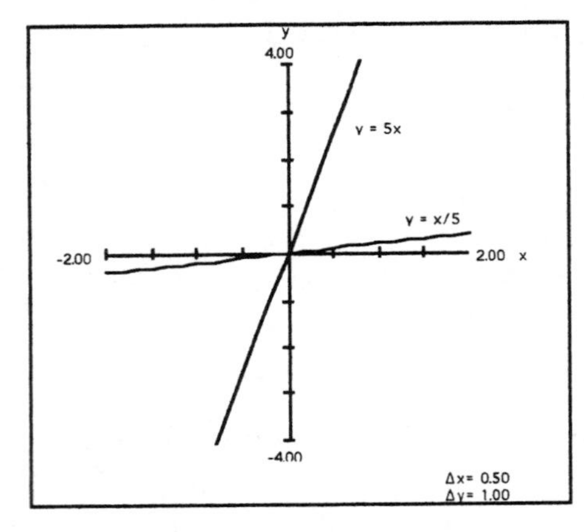

Graphs of $t(x) = 1/x$ and $f(x) = 5/x$

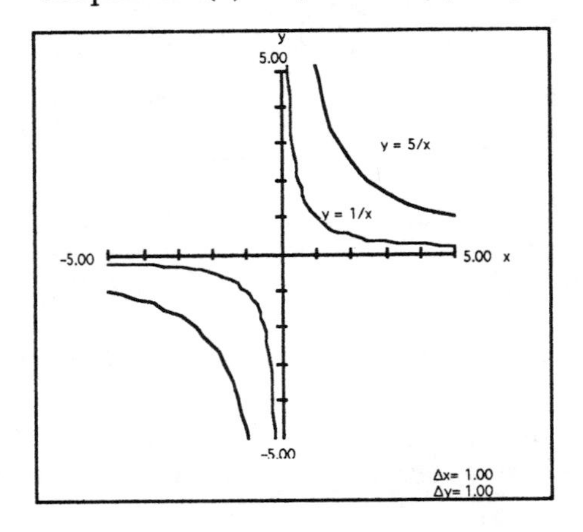

45. a. The graph of $y = x^2$ decreases for $x < 0$ and increases for $x > 0$ (as can be seen in the diagram on the left).

The graph of $y = x^{-2}$ increases for $x < 0$ and decreases for $x > 0$ (as can also be seen from that diagram).

b. The two graphs intersect at (-1,1) and (1,1).

c. As x approaches $\pm\infty$, the graph of $y = x^2$ approaches ∞. As x approaches $\pm\infty$, the graph of $y = x^{-2}$ approaches 0.

47. a. $f(x) = 2x$ and $g(x) = 4x^2$ intersect at $x = 0$ and at $x = 1/2$ since $4x^2 - 2x = 2x(2x-1)$.

b. $f(x) = 4x^2$ and $g(x) = 4x^3$ intersect at $x = 0$ and at $x = 1$ since $4x^2 - 4x^3 = 4x^2(1-x)$.

c. $f(x)=x^{-2}$ and $g(x)=4x^2$ intersect at $x=\pm(1/4)^{1/4} = \pm\sqrt{1/2}$ since $x^{-2} - 4x^2 = (1 - 4x^4)/x^2$

d. $f(x) = x^{-1}$ and $g(x) = x^{-2}$ intersect at $x = 1$ since $f(x) - g(x) = x^{-1} - x^{-2} = (x-1)/x^2$.

e. $f(x) = 4x^{-2}$ and $g(x) = 4x^{-3}$ intersect at $x = 1$. (Solve $4x^{-2} = 4x^{-3}$.)

Exercises for Section 7.7

49. The graph given in the text suggests an exponential decay function. It is nearly a straight line graph and the vertical scale is logarithmic.

51. a.

Species	adult mass in grams	egg mass in grams	egg/adult ratio
ostrich	113,380.0	1700.0	0.015
goose	4,536.0	165.4	0.036
duck	3,629.0	94.5	0.026
pheasant	1,020.0	34.0	0.033
pigeon	283.0	14.0	0.049
hummingbird	3.6	0.6	0.167

Your answer may be different. Notable is the fact that the egg/adult mass ratio for the humming bird is very high and it is very low for the ostrich. The other ratios are not far apart from each other.

b. The plots are respectively

i. linear x *vs.* linear y

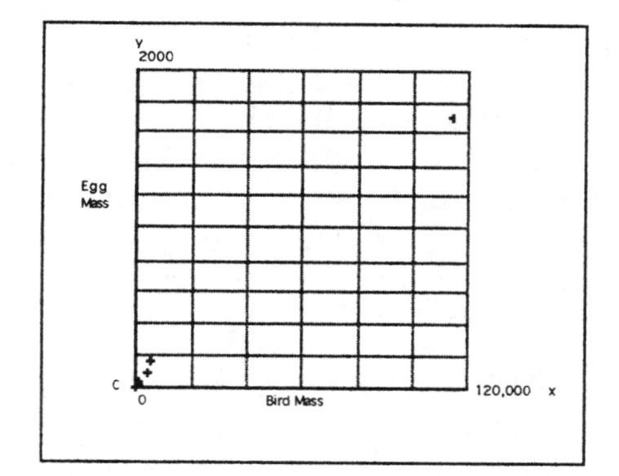

ii. linear x vs. log y

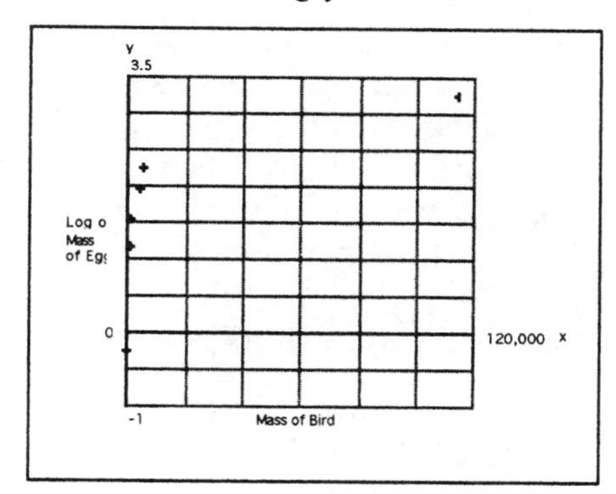

iii. log x vs log y

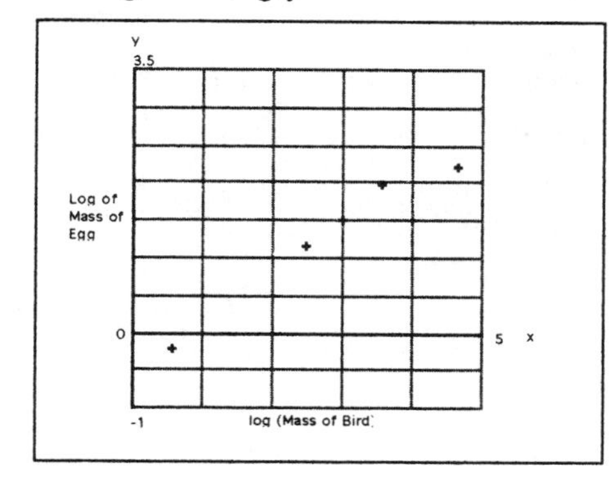

c. The log *vs.* log scatter diagram (on the left) is, of course, the most linear. The best straight-line fit equation for this is: log(y) = log(0.1918) + 0.7719•log(x) or, in linear equation form: Y = -0.717+ 0.7719X where Y = log(y) and X = log(x).

d. Regrouping, using the laws of logarithms, we get: $\log(y) = \log(0.1918 \bullet x^{0.7719})$. Thus, $y = 0.1918 \bullet x^{0.7719}$, where x = mass of the adult and y is the corresponding mass of the egg, both measured in grams.

e. If $x = 12.6$ kg. for the weight of an adult turkey, then $y \approx 1.36$ grams is the predicted weight of its egg.

f. If the egg weighs 2 grams, then the adult bird is predicted, by reverse solving, to have an adult weight ≈ 20.848 grams.

53. a. Since ℓ_1 and ℓ_2 are parallel lines they have the same slope but different vertical intercepts. Thus their equations can be written as $\log(y) = mx + \log(b_1)$ and $\log(y) = mx + \log(b_2)$, with $\log(b_1) \neq \log(b_2)$. Solving each for y in terms of x we have: $y = b_1 \bullet 10^{mx}$ and $y = b_2 \bullet 10^{mx}$ with $b_1 \neq b_2$. Thus they have the same power of 10 but differ in their y-intercepts.

b. The equation of ℓ_3 is $\log(y) = m \bullet \log(x) + \log(b_3)$ and the equation of ℓ_4 is $\log(y) = m \bullet \log(x) + \log(b_4)$, with $\log(b_3) \neq \log(b_4)$. Solving for y in terms of x in each gives: $y = b_3 x^m$ and $y = b_4 x^m$ with $b_3 \neq b_4$. Thus they have the same power of x but differ in their constant multipliers.

Exercises for Section 7.8

55. a. The population density is directly proportional to the -2.25 power of the length of the organism. Here is a derivation. Let x be the length of the organism measured in meters and let p be the population density of that organism, *i.e.*, the number of individual organisms per square kilometer. Then, since, the scale on both axes of the graph is logarithmic and the graph is a straight line, the relationship between population density and length of an organism is: $\log(p) = k - 2.25 \bullet \log(x)$. Since we can write $k = \log(c)$ for some positive c, we can combine and get $\log(p) = \log(c \bullet x^{-2..25})$ or $p = c \bullet x^{-2.25}$.

b. The population density of the organism is directly proportional to the -0.75 power of the body mass of the organism. Here is a derivation. In this part m measures body mass in kilograms and p measures population density in the same units as in part **a.** We argue in a similar fashion and obtain that $\log(p) = \log(d) - 0.75 \bullet \log(m)$ or $p = d \bullet m^{-0.75}$.

c. From **a.** and **b.** we have that $d \bullet m^{-0.75} = c \bullet x^{-2..25}$ or that $m = (d/c)^{4/3} \bullet x^3$. Thus, mass is directly proportional to the cube of the length.

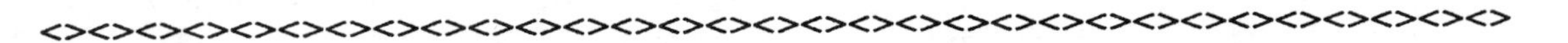

Exercises for Section 8.1

1. **a.** 1/8, -1/8 **b.** 1/2, -1/2 **c.** -1/2, 1/2 **d.** -32, 32

3. **a.** $j(x) = 3x^5 + x^2 + x - 1$; $k(x) = 3x^5 - x^2 + x + 1$; $l(x) = 3x^7 - 3x^5 + x^3 - x$
b. $j(2) = 101$; $k(3) = 724$ $l(-1) = 0$

5. **a.** $y = 2x - 3$ goes with Graph 1: linear, positive slope, y-intercept at -3.

b. $y = 2 - x$ goes with Graph 4: linear, negative slope; y-intercept at 2.

c. $y = 3(2^x)$ goes with Graph 3: exponential; increasing; y-intercept at 3.

d. $y = (x^2 + 1)(x^2 - 4)$ goes with Graph 2: 2 zeros at æ2, 4th degree, *etc.*

7. The graphs provided here are based upon numerical substitution and can be used as guides for your own rougher and geometrically based sketches.

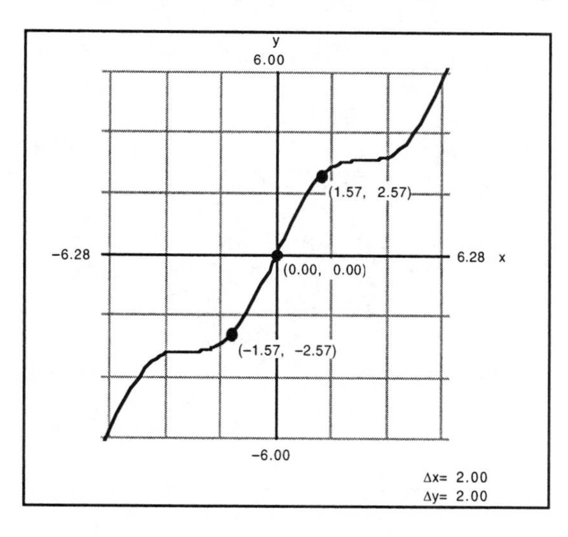

a.

x	$y_1 = x$	$y_2 = \sin x$	$y_1 + y_2$
0.0	0.0	0.0	0.0
1.6	1.6	1.0	2.6
-1.6	-1.6	-1.0	-2.6
...	...	...	...

(Actually $y_1(x) = x$ and $y_2(x) = \sin x$ but you will eyeball the y values and thus the actual shape of the graph drawn will probably be a bit different.)

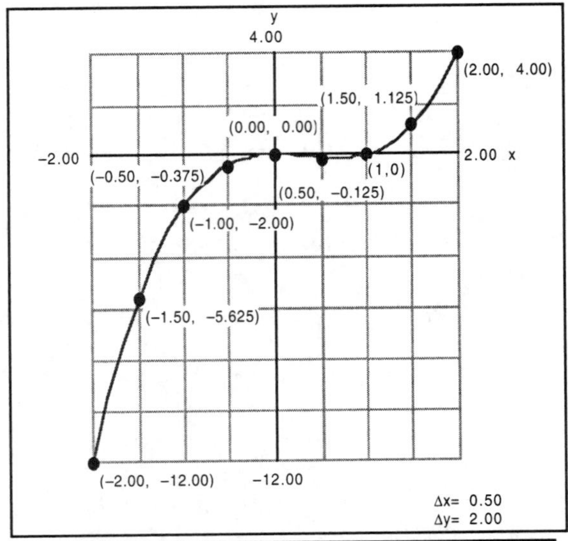

b.

x	$y_1 = -x^2$	$y_2 = x^3$	y_1+y_2
0.0	0.00	0.000	0.000
0.5	-0.25	0.125	1.000
1.0	-1.00	1.000	0.000
1.5	-2.25	3.375	1.125
2.0	-4.00	8.000	4.000
-0.5	-0.25	-0.125	-1.500
-1.0	-1.00	-1.000	-2.000
-1.5	-2.25	-3.375	-5.625
-2.0	-4.00	-8.000	-12.000

The graph of $y = -x^2 + x^3$ for $-2 \leq x \leq 2$ is given on the left; since $y = x^2 \bullet (x-1)$ it has 0 and 1 as its zeros.

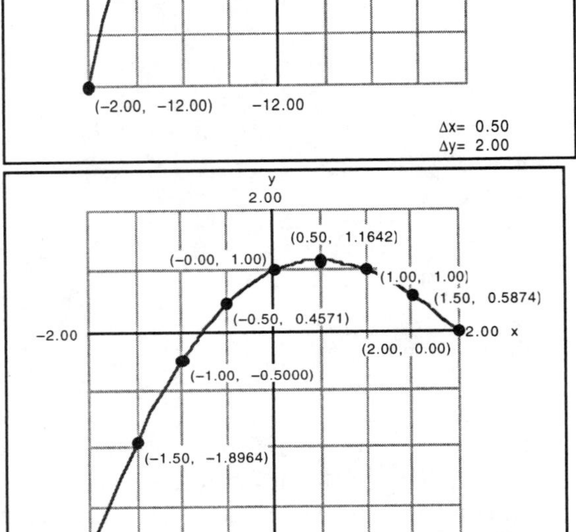

c.

x	$y_1 = 2^x$	$y_2 = -x^2$	y_1+y_2
2.0	0.2500	-4.00	-3.7500
-1.5	0.3536	-2.25	-1.8964
-1.0	0.5000	-1.00	-0.5000
-0.5	0.7071	-0.25	0.4571
0.0	1.0000	0.00	1.0000
0.5	1.4142	-0.25	1.1642
1.0	2.0000	-1.00	1.0000
1.5	2.8284	-2.25	0.5874
2.0	4.0000	-4.00	0.0000

Its graph is given on the left with the points in the table marked on it.

9. a. $8701.5 = 8n^3 + 7n^2 + 0n + 1n^0 + 5n^{-1}$.

b. $239 = 2n^2 + 3n + 9n^0$; if $n = 2$, this polynomial would have the value 23.

c. The number written in base 2 as 11001 evaluates to 25 when written in base 10 notation, since $1 \bullet 2^4 + 1 \bullet 2^3 + 0 \bullet 2^2 + 0 \bullet 2^1 + 1 \bullet 2^0 = 25$.

d. Here is one way to find the base two equivalent: find the highest power of 2 in 35. This is $2^5 = 32$. Subtracting that from 35 leaves 3 which is easily written as 2 + 1. Thus 35, in base 10, can be written as $1 \bullet 2^5 + 0 \bullet 2^4 + 0 \bullet 2^3 + 0 \bullet 2^2 + 1 \bullet 2^1 + 1 \bullet 2^0$ and this is 100011 in base 2.

11.

Function	x-intercepts	degree
$y = 2x + 1$	-0.5	1
$y = x^2 - 3x - 4$	-1, 4	2
$y = x^3 - 5x^2 + 3x + 5$	-0.709, 1.806, 3.903	3
$y = 0.5x^4 + x^3 - 6x^2 + x + 3$	-4.627, -0.610, 0.916, 2.320	4

One might be tempted to conclude that a polynomial of degree n has n distinct real zeros. But be careful!

13. a. $y = 3x^3 - 2x^2 - 3$ has only one real zero at $x \approx 1.28$.

b. $y = x^2 + 5x + 3$ has two real zeros: $x = -2.5 \pm 0.5\sqrt{13} \approx -4.30$ or -0.70

Graph for **#13 a.**

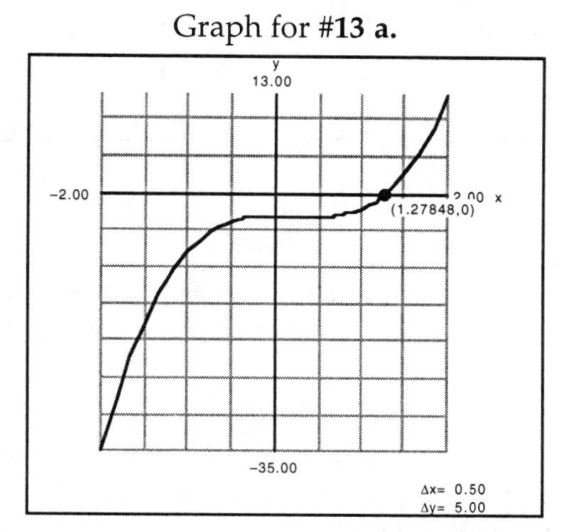

Graph for **#13 b.**

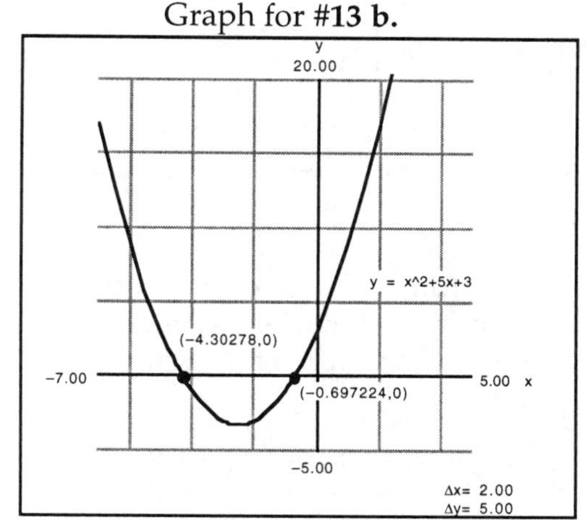

15. a. When the degree of a polynomial is odd:

If its leading coefficient is positive then for large negative values of x the y values of this polynomial are negative and for large positive values of x, the y values of this polynomial are positive. The graph must therefore cross the x axis to go from the negative to the positive y values, since the graph of a polynomial is one piece. Thus it must have at least one real zero.

When its leading coefficient is negative, the roles of positive and negative in the previous paragraph are reversed but the conclusion is the same, namely, it must have at least one real zero.

b. **i.** $y = x^2 + 1$ is an example of a polynomial of degree 2 that has no real zeros. Its graph is below on the left.

ii. $y = -1 - x^4$ is another example of a polynomial of degree 2 that has no real zeros. Its graph is below on left.

c. **i.** $y = x^3$ is an example of a polynomial of degree 3 that has exactly one real zero. Its graph is given below in the middle.

ii. $y = (x + 1)(x - 3)(x - 5)$ is an example of a polynomial of degree 3 that has exactly 3 real zeros. Its graph is also below in the middle.

d. The graph of $y = (x^4 - 1)$ has exactly two real zeros: 1 and -1. It is displayed below on the right.

Graphs for **#15 b.**

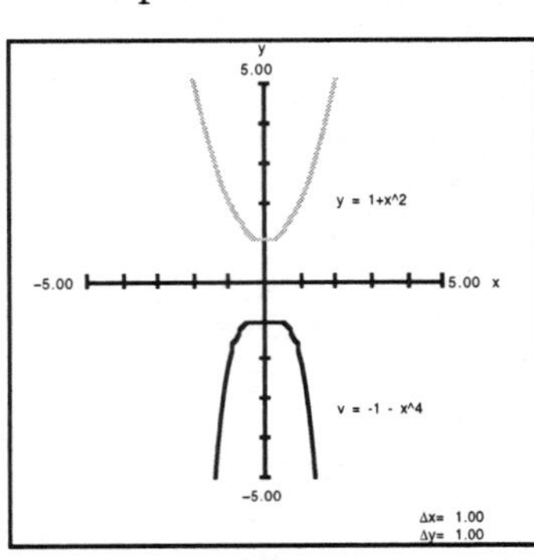

Graphs for **#15 c.**

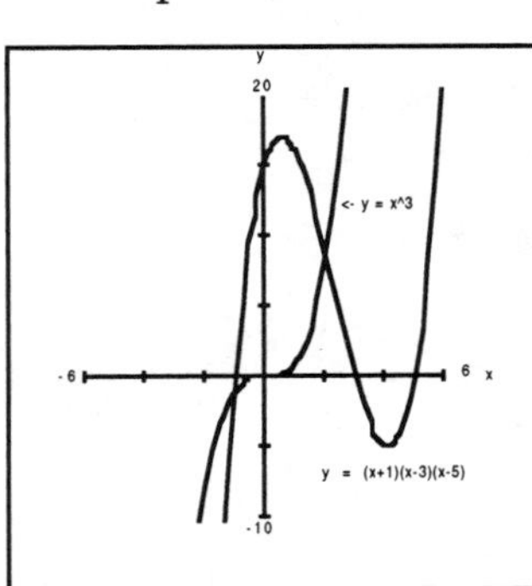

Graph for **#15d.**

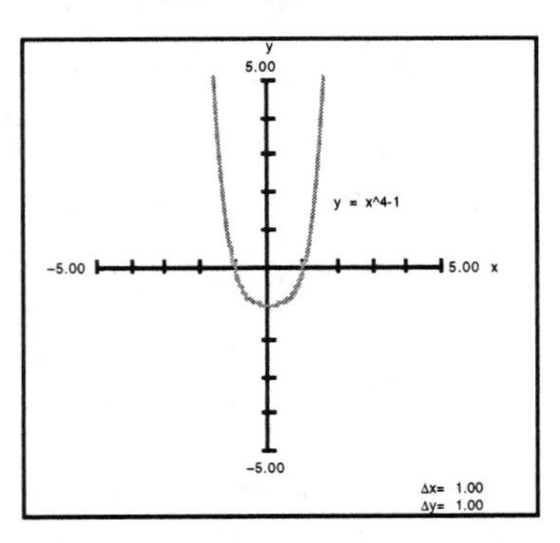

17. a. If $f(x) = a \bullet x^{2k}$, then $f(-x) = a(-x)^{2k} = ax^{2k} = f(x)$

b. If $f(x) = a \bullet x^{2k+1}$, then $f(-x) = a(-x)^{2k+1} = -ax^{2k} = -f(x)$.

c. Even functions have graphs that are symmetric about the y axis. Odd functions have graphs that are symmetric about the origin. An inspection of the graphs of $y = x^3$ and of x^2, for example, will show this.

d. **i.** $f(-x) = (-x)^4 + (-x)^2 = x^4 + x^2 = f(x)$ -- this is an even function.
ii. $u(-x) = (-x)^5 + (-x)^3 =(-x^5) + (-x^3) = -(x^5+x^3) = -u(x)$ -- this is an odd function.
iii. $h(-x) = (-x)^4 + (-x)^3 = x^4 - x^3 \neq h(x)$ and $\neq -h(x)$ -- neither even nor odd.
iv. $g(-x) = 10 \bullet 3^{-x} \neq g(x)$ and $\neq -g(x)$ -- neither even nor odd.

e. The graphs of even functions are symmetric with respect to the y axis and the graphs of odd functions are symmetric with respect to the origin as the graphs of the functions in **d. i.** and **d. ii.** will show.

Exercises for Section 8.2

19. a. The graphs of these are all parabolas facing up and with vertex at (0,0). But the larger the coefficient, the faster the y values grow as x moves away from the origin in either direction. (See the graphs below on the left.)

b. The graphs of these functions are all parabolas. Two have their vertex at (0,4) and one has the vertex at (0, -4). The first and last are mirror images of each other with respect to the x axis. The first and 2nd are mirror images of each other with respect to the graph of $y = 4$. (See the graphs below in the middle.)

c. The y values of all three increase as x increases when $x > 0$. Their relative rates of increase vary; the linear grows more rapidly than the quadratic for $0 < x < 1$ and more rapidly than the exponential until somewhere between $x = 1.5$ and

1.6; the quadratic grows more rapidly than the exponential until x reached approximately 3.3. where its rate of growth then becomes the fastest.

Graphs for **#19. a.** Graphs for **#19. b.** Graphs for **#19. c.**

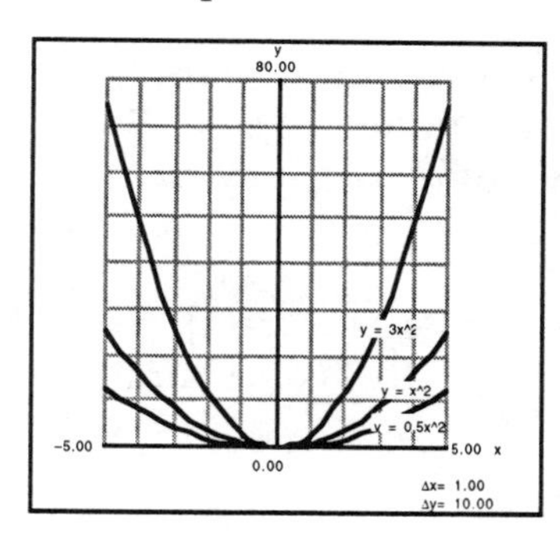

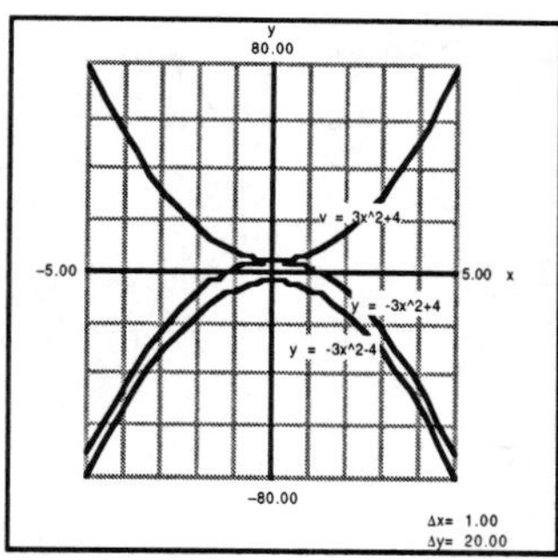

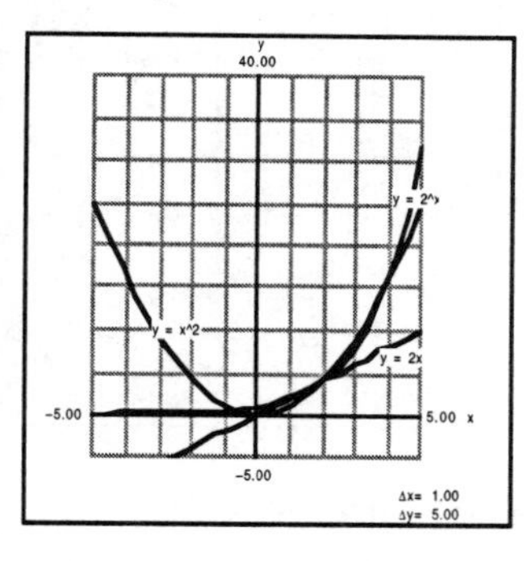

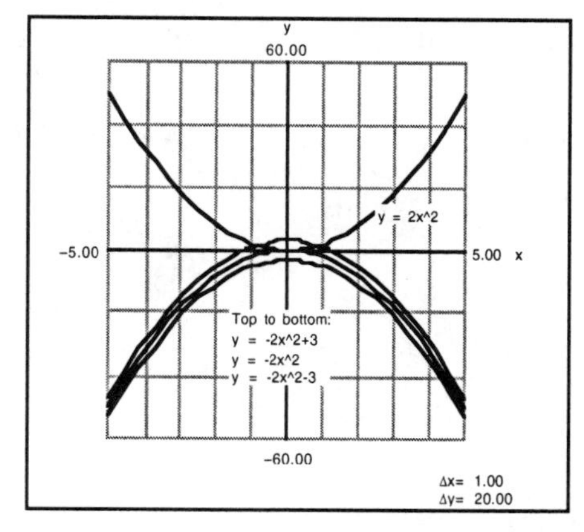

21. The graphs of the four functions are given in with their labels the diagram to the left.

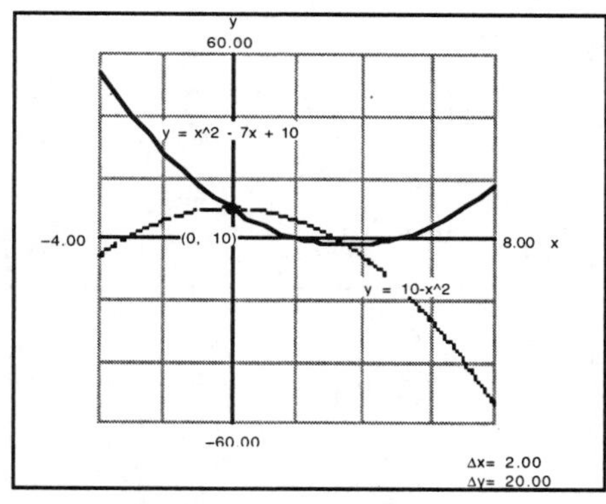

23. Any quadratic of the form $y = ax^2 + bx + 10$ will have a y-intercept of 10. If a is positive then the graph will turn up and if a is negative the graph will turn down. The value of b is not relevant here. The graph of $y = x^2 - 7x + 10$, for example, is turned up and has two y-intercepts, one at 5 and the other at 2. (Your answer will probably be different.) The graph of $y = 10 - x^2$ will turn down and its two zeros are $x = \pm\sqrt{10}$. The graphs are given in the diagram to the left

25. a. $A(x) = x^2 + (20-x)^2$ **b.** Its domain is $0 < x < 20$

27. a. 50 meters by 50 meters **b.** P/4 meters by P/4 meters

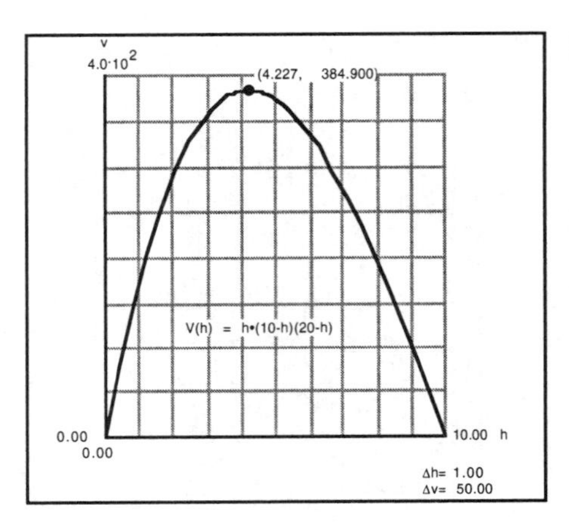

29. The volume is given by the formula:

$$V(h) = (20-h)\bullet(10-h)\bullet h$$

with $0 < h < 10$.

An inspection of the graph shows that the maximum volume occurs at $h \approx 4.2$ feet. The corresponding volume is approximately 384.9 cu.ft.

(The dimensions of the box are approximately 4.2 ft. by 5.8 ft. by 15.8 ft. That is a rather large box!)

31. a. $W = 1600\bullet4\bullet100/120 = 5333.33$ lbs.

b. $W = 1600\bullet10\bullet16/120 = 2133.33$ lbs. It has been reduced to 40% of what it should be.

c. If L is doubled, then W is halved and equals 2666.67

d. $1600\bullet4\bullet d^2/192 = 1600\bullet4\bullet100/120$ gives $d^2 = 160$ or $d \approx 12.65$ in.

Exercises for Section 8.3

33. a. $6t^2 - 7t - 5 = 0$ give $t = (7 \pm \sqrt{169})/12 = (7 \pm 13)/12$ or $t = 5/3$ or $-1/2$.
[Note that $6t^2 - 7t - 5 = (3t - 5)(2t + 1)$.]

b. $9x^2 - 12x + 4 = 0$ gives $x = (12 \pm \sqrt{0})/18$ or $x = 2/3$.
[Note that $9x^2 - 12x + 4 = (3x - 2)^2$]

c. $3z^2 - z - 9 = 0$ gives $z = (1 \pm \sqrt{109})/6$ or $z \approx 1.91$ or -1.57.

d. $x^2 + 6x + 7 = 0$ gives $x = (-6 \pm \sqrt{8})/2$; thus $x \approx -1.59$ or -4.41.

35. a. $y = x^2 - 5x + 6$ has zeros at 2 and 3

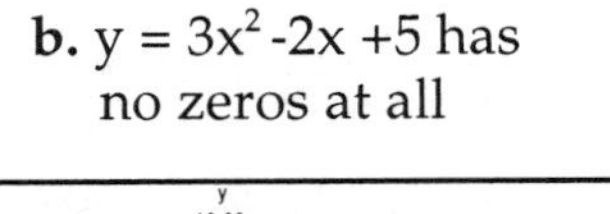
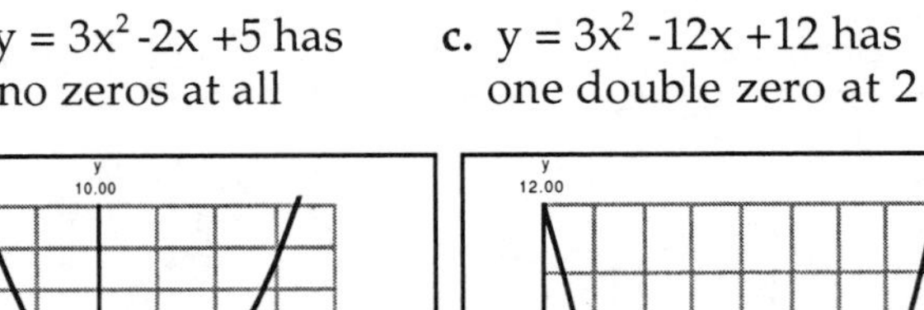

b. $y = 3x^2 - 2x + 5$ has no zeros at all

c. $y = 3x^2 - 12x + 12$ has one double zero at 2

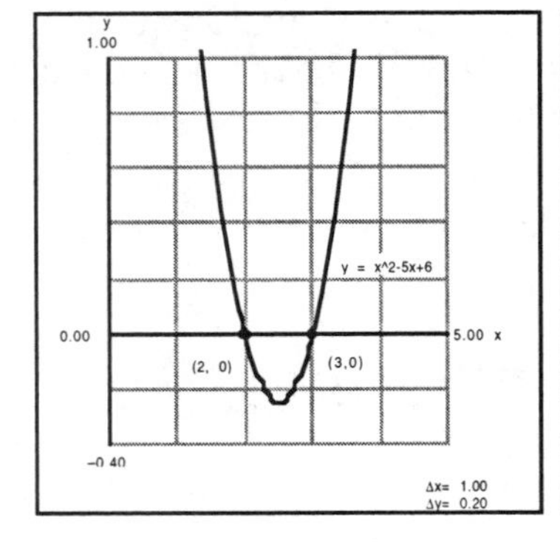

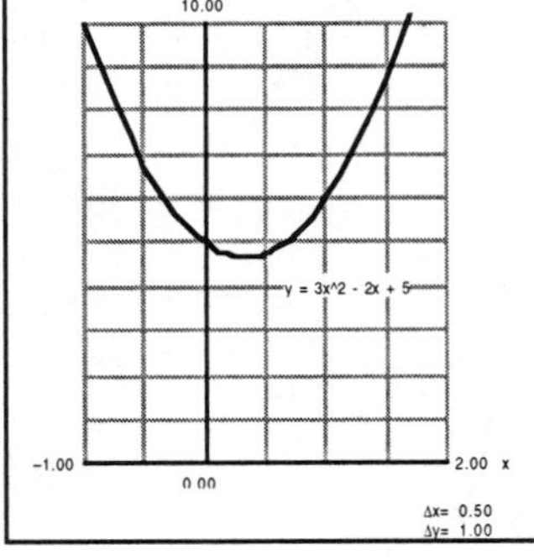

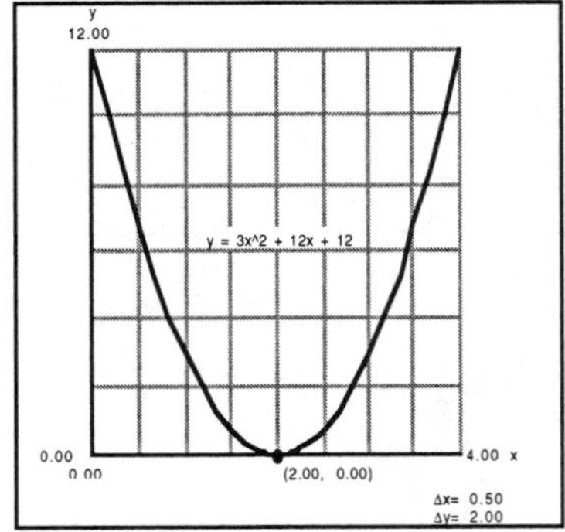

37. a. The y-intercept is 1 (obtained by plugging in x = 0). Using the quadratic formula gives the x-intercepts as $(-2 \pm \sqrt{16})/6$; thus x = 1/3 or -1; factoring gives y = (3x-1)(x+1).

b. The y-intercept is 11. Using completed square form to get the x-intercepts gives $x - 2 = \pm\sqrt{1/3}$ or $x = 2 \pm \sqrt{1/3}$ and thus $x \approx 1.42$ or 2.58

39. a. $0 = x^2 - 9 = (x - 3)(x + 3)$; thus x = 3 or -3
b. $0 = x^2 - 4x = x(x - 4)$; thus x = 0 or 4.
c. $0 = 3x(3x - 25)$; thus x = 0 or 25/3
d. $0 = x^2 + x - 20 = (x + 5)(x - 4)$; x = -5 or 4.
e. $0 = 4x^2 - 12x + 9 = (2x-3)^2$; x = 3/2
f. $0 = 3x^2 - 13x-10 =(3x+2)(x-5)$; x = 5,-2/3
g. $0 = x^2 + 4x + 4 = (x+2)^2$; x = -2
h. $0=2x^2 -5x -3 = (2x+1)(x-3)$; x = -0.5, 3

41. **a.** y = (x+4)(x+2)

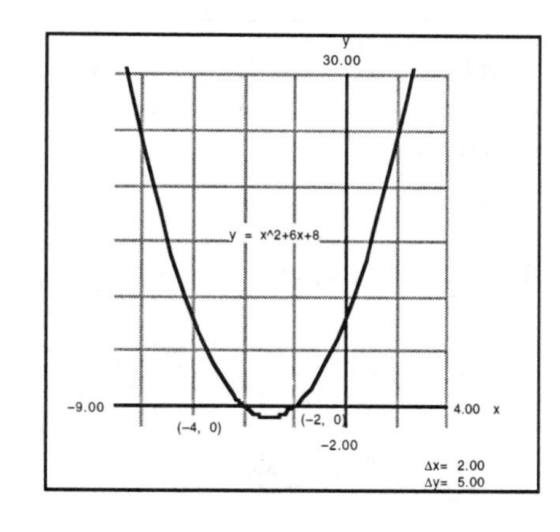

41. b. z = 3(x-3)(x+1)

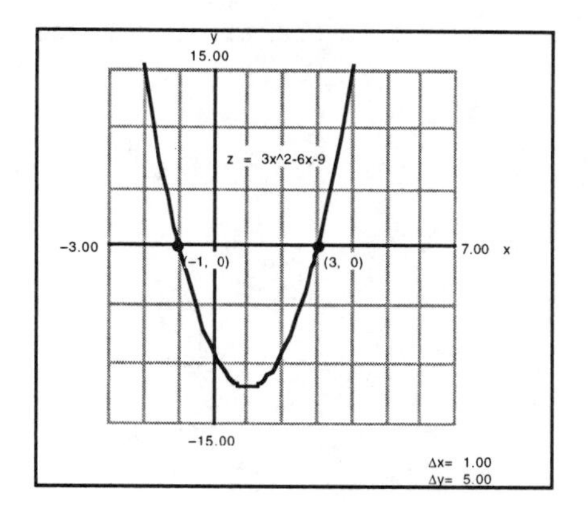

41. c. f(x) = (x-5)(x+2)

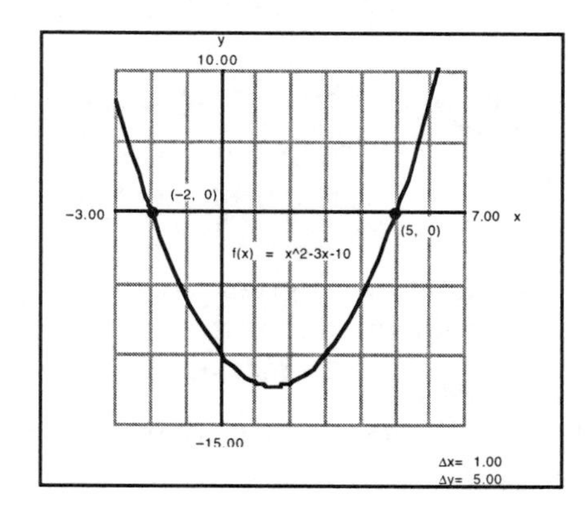

41. d. w = (t-5)(t+5)

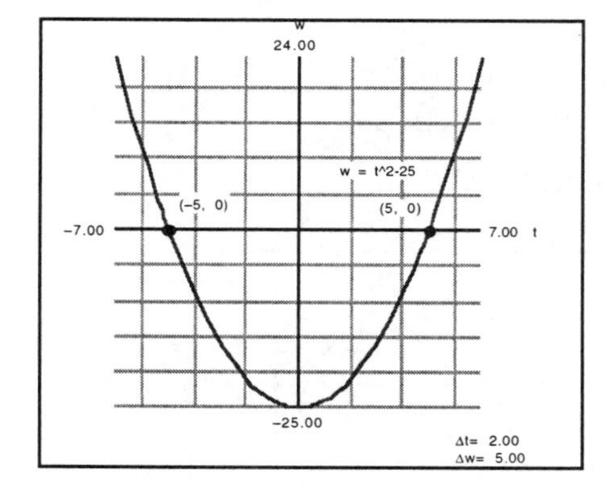

41. e. $r = 4(s-5)(s+5)$

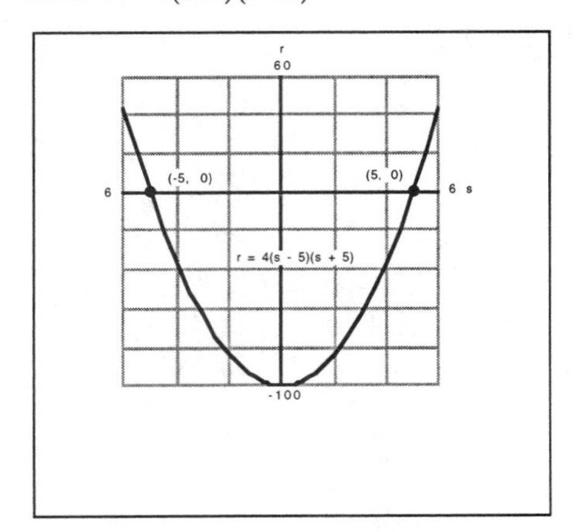

41. f. $g(x) = (x+1)(3x-4)$

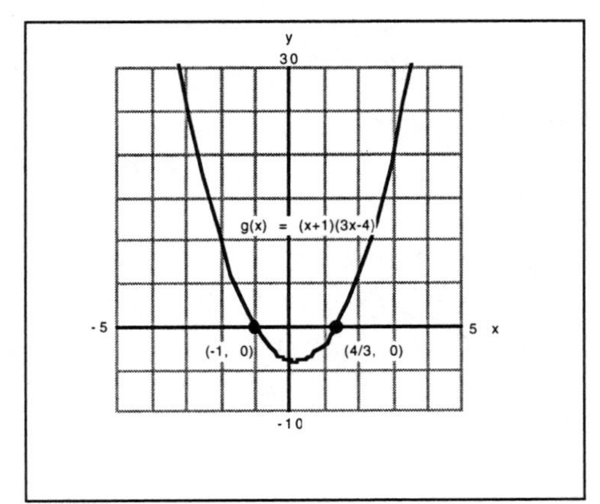

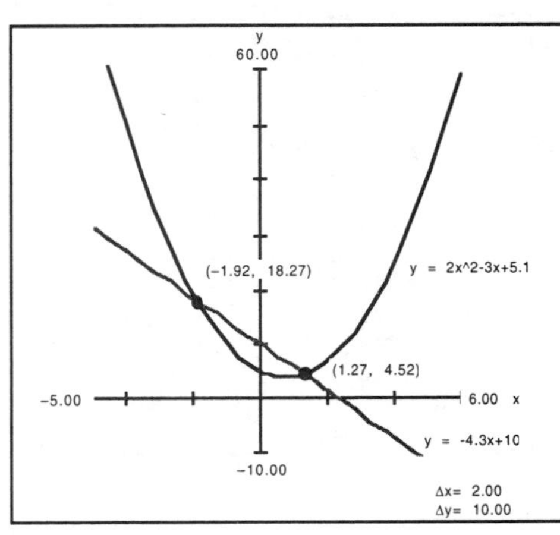

43. a. Algebraically one finds the values of x that satisfy both equations as follows:

$2x^2 - 3x + 5.1 = -4.3x + 10$ or

$2x^2 + 1.3x - 4.9 = 0$ or

$$x = \frac{-1.3 \pm \sqrt{1.3^2 - 4(2)(-4.9)}}{2 \cdot 2} \quad \text{or}$$

$$= \frac{-1.3 \pm \sqrt{40.89}}{4} \quad \text{or}$$

$$= \frac{-1.3 \pm 6.39}{4} \approx 1.27 \text{ or } -1.92$$

b. The graphs are given in the diagram on the right; they show the points of intersection at x = -1.92, y = 18.27 and at x = 1.27, y = 4.52

Exercises for Section 8.4

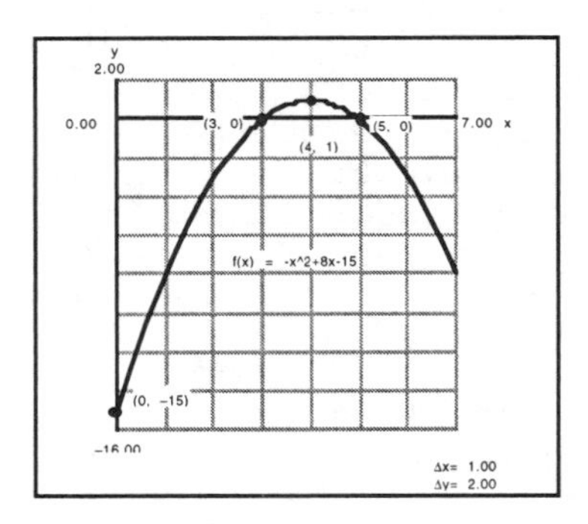

45. The graph of the function is given in the diagram on the left, with coordinates of the y-intercept, x-intercepts and vertex clearly marked.

a. The formula can be factored into $y = -(x-3)(x-5)$ and thus the zeros are 3 and 5.

b. $f(0) = -15$ gives the y-intercept.

c. The vertex is at $x = -b/(2a) = -8/(-2) = 4$

47. The maximum profit is had when $x = -20/(2\bullet -0.5) = 20$; the maximum profit is 430 thousand dollars.

49. The maximum occurs when $x = -48/-6 = 8$ computers and the revenue from selling 8 will be 192 million dollars.

51. a. At $t = 0$, $h = 4$ ft.

b. By the quadratic formula, $h = 0$ when $t = (-50 \pm \sqrt{2756})/(-32) \approx 3.20$ or -0.08 and the latter is rejected because h is not defined for negative values of time.
c. $h = 30$ ft. at $t = 0.659$ sec. and at $t = 2.466$ sec. (via quadratic formula); it is never 90 feet high since the discriminant in the resultant quadratic formula is -3004.

d. It reaches its maximum height at $t = -50/(-32) \approx 1.56$; and the height at that moment is approximately 43.06 ft.

53. We are given that $2W + L = 1$. Thus $L = 1 - 2W$ and thus the area formula is $A = W(1-2W) = W - 2W^2$. Note this is at its maximum when W is at the vertex, i.e., when $W = -1/[2\bullet(-2)] = 1/4$. At this value of W we have that $L = 1/2$. Thus, Dido was correct.

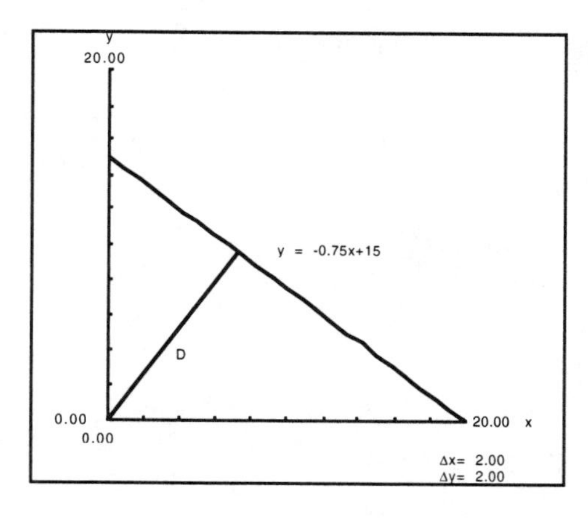

55. a. The graph is illustrated on the left, with x and y marked off in miles, and with a typical point (x,y) marked on the highway along with the straight line to that point.

b. The highway goes through the points (0,15) and (20,0) and thus has the equation $y = -0.75x + 15$.

c. and **d.** The distance squared

$$D = x^2 + y^2 = x^2 + (15 - .75x)^2$$
$$= 1.5625x^2 - 22.5x + 225$$

e. Since $D = 1.5625x^2 - 22.5x + 225$, the minimum occurs at the vertex, which is at $x = 22.5/(2\bullet 1.5625) = 7.2$. The minimum for D is 144 and thus the minimum distance is $\sqrt{144}$ or 12 miles.

f. The coordinates of the point of shortest distance from (0,0) are (7.2, 9.6)

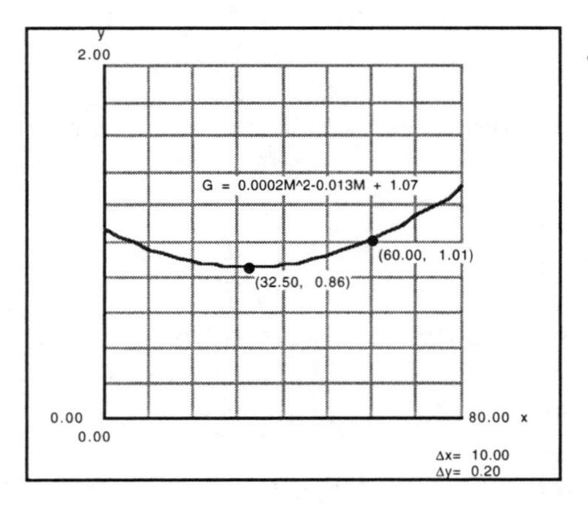

57. a. and **b.** The graph is to the left. The minimum gas consumption rate suggested by the graph occurs when M equals 32.5 mph. and it is approximately 0.85 gph. Computation on a calculator gives the same M but the gas consumption rate is 0.86 when rounded off.

c. In 2 hours 1.72 gallons will be used and you will have traveled 65 miles.

d. If M = 60 mph then G = 1.01 gph. It takes 1 hour and 5 minutes to travel 65 miles at 60 mph and one will have used 1.094 gallons.

e. Clearly traveling at the speed that supposedly minimizes the gas consumption rate does not conserve fuel if the trip lasts only 2 hrs.

f. and **g.**

mph	gph	gpm	mpg
0	1.07	----------	0.0
10	0.96	0.09600	10.4
20	0.89	0.04450	22.5
30	0.86	0.02867	34.9
40	0.87	0.02175	46.0
50	0.92	0.01840	54.3
60	1.01	0.01684	59.4
70	1.14	0.01629	61.4
80	1.31	0.01638	61.1

For **f.** we have:

$$G/M = (0.0002M^2-0.013M+1.07)/M.$$

Its graph is below on the left. Eyeballing gives the minimum y at M ≈ 73 mpg.

For **g.** we have:

$$M/G = M/(0.0002M^2 -0.013M+1.07).$$

Its graph is given on the next page on the right. Eyeballing gives the maximum y at the same M. This is expected. since max = 1/min

Graph for #57. f. $y = G/M$

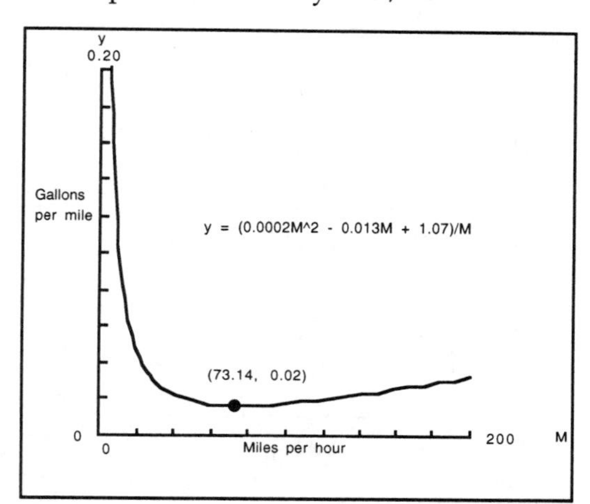

Graph for #57 g. $y = M/G$

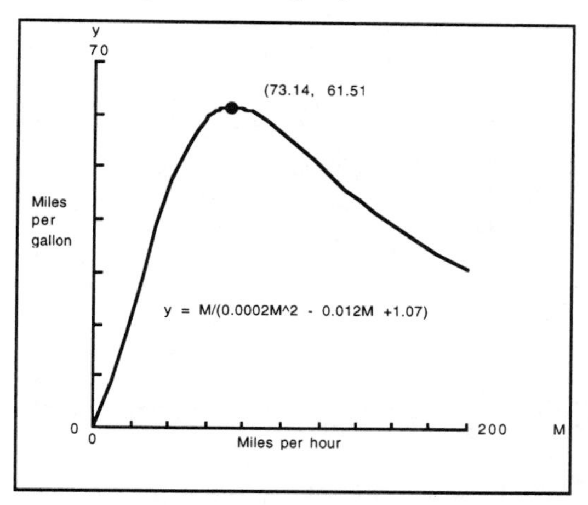

59. In general: the a-h-k form is the easier to use to find the vertex and the x-intercepts; the a-b-c form is easier to use to find the y-intercept from.

a. a-b-c form: $y_1 = 2x^2 - 3x - 20$ a-h-k form: $y_1 = 2(x - 0.75)^2 - 21.125$.
a-b-c form: $y_2 = -2x^2 + 4x - 5$ a-h-k form: $y_2 = -2(x-1)^2 - 3$
a-b-c form: $y_3 = 3x^2 + 6x + 3$ a-h-k form: $y_3 = 3(x+1)^2 + 0$
a-b-c form: $y_4 = -2x^2 + 2x + 12$ a-h-k form: $y_4 = -2(x - 0.5)^2 + 12.5$

b. The vertex for y_1 is at (0.75, -21.125); its y-intercept is -20; its x-intercepts are at $x = -2.5$ and $x = 4$. Its graph is given on the top left on the next page.

The vertex for y_2 is at (1,-3); its y-intercept is -5; it has no x-intercepts since $(x-1)^2$ cannot be equal to -3/2 and its discriminant is -24. Its graph is given on the top right on the next page

The vertex for y_3 is at (-1,0); its y-intercept is 3 and its x-intercept is at $x = -1$ only. Its graph is given on the left on the next page.

The vertex for y_4 is at (0.5, 12.5); its y-intercept is 12; its x-intercepts are at $x = -2.$ and $x = 3$. Its graph is given below on the right.

Graph of y_1

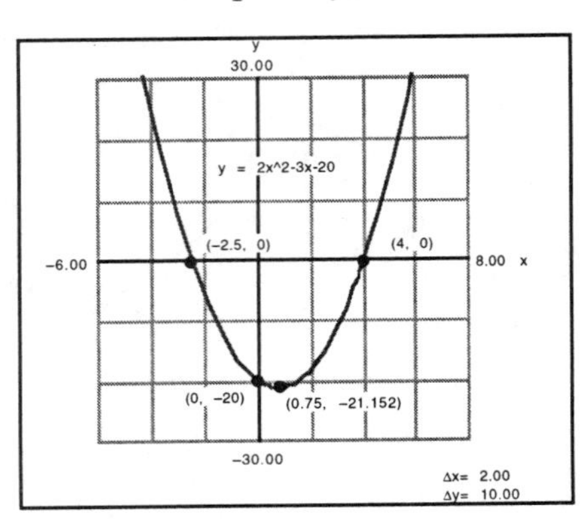

Graph of y_2

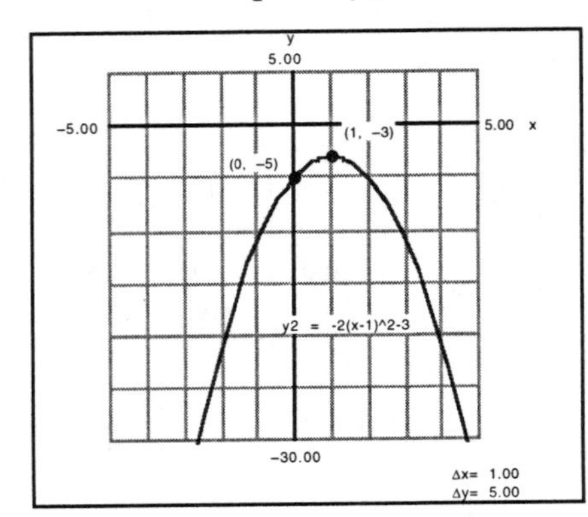

Graph of y_3

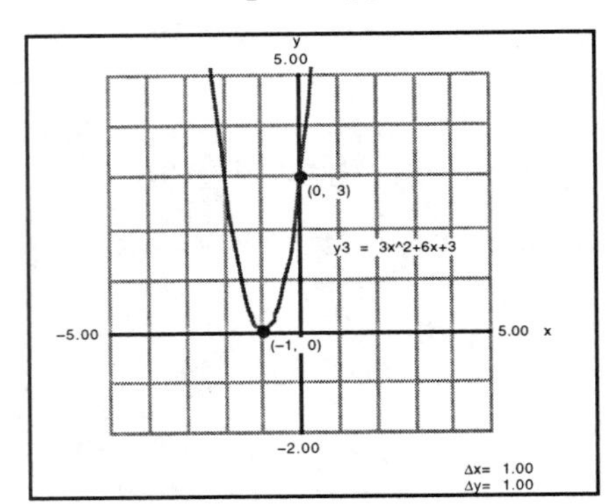

Graph of y_4

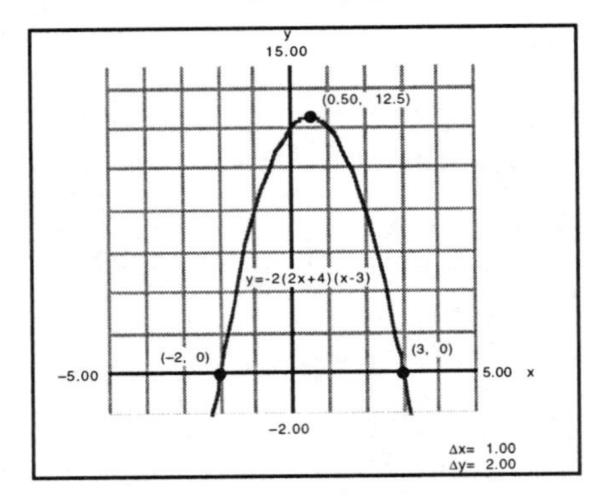

61. a. $y = a(x-2)^2 + 4$ and $7 = a(1-2)^2 + 4$ gives $a = 3$ and thus $y = 3(x-2)^2+4 = 3x^2 -12x+16$

b. If $a > 0$, the graph of $y = a(x-2)^2 - 3$ faces up; it faces down if $a < 0$.

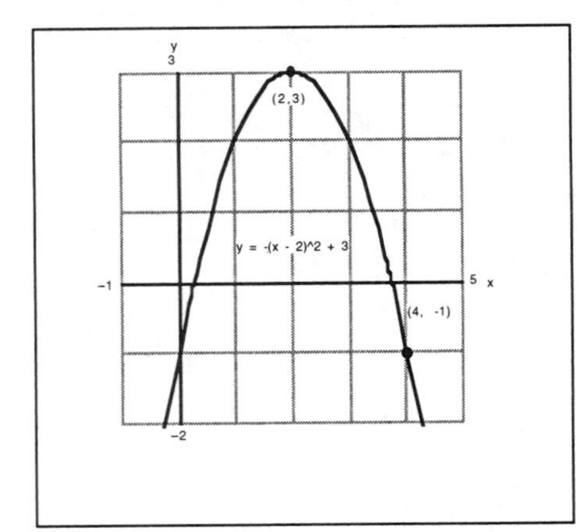

63. Derivation: from the data we have:

$$y = a(x - 2)^2 + 3; \text{ and}$$
$$-1 = a(4-2)^2 + 3$$

This implies that $a = -1$;

Thus $y = -(x - 2)^2 + 3$ is its equation.

Check: $-(4-2)^2 + 3 = -4 + 3 = -1$

This checks out with the graph on the left.

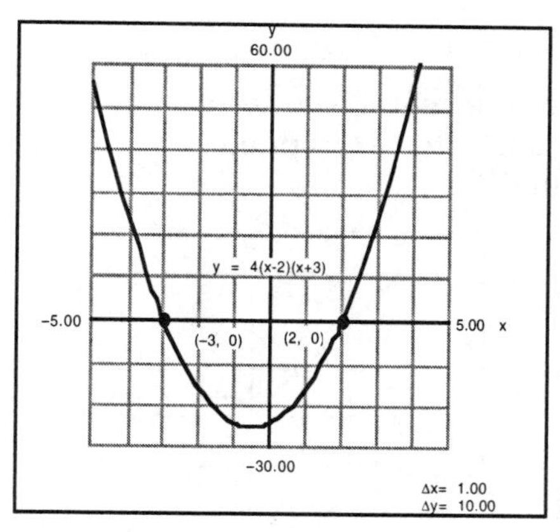

65. The equation is $y = 4(x-2)(x+3) = 4x^2 + 4x - 24$.

Reasons: The coefficient of x^2 is always a constant multiplier of the factored form and the two zeros are given.

This checks out with the graph on the left.

67. a. The focus is at $\left(-\dfrac{b}{2a}, c + \dfrac{1-b^2}{4a}\right)$

b. i. (-3/2, 0) and = 1/4

c. Graph of $y = x^2 + 3x + 2$ showing its focus

ii. (1, 3.125) and 0.125
Graph of $y = 2x^2 - 4x + 5$ showing its focus

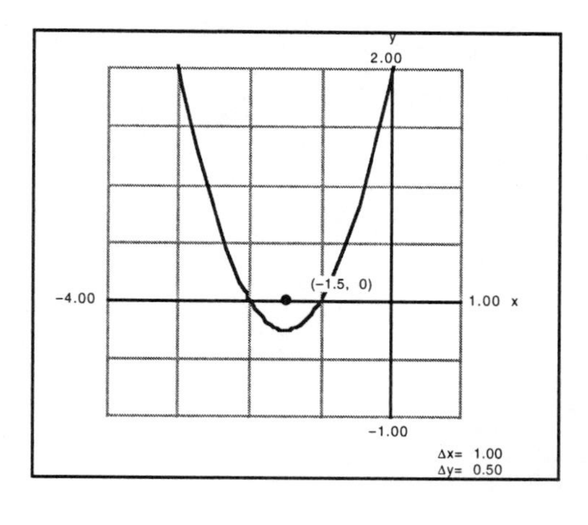

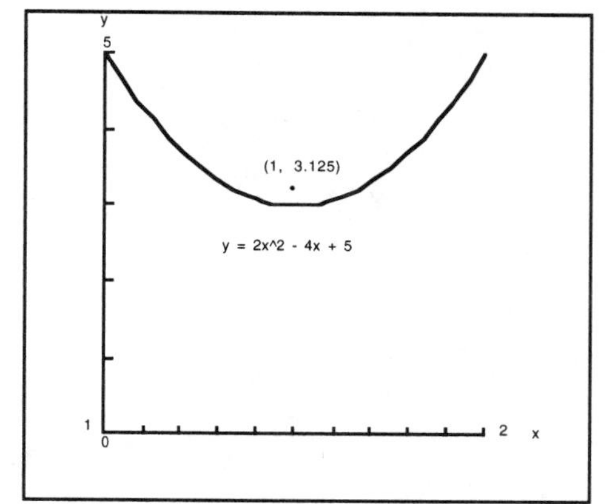

d. The smaller in absolute value is a, the more open is the parabola and thus the larger is the focal length, since the focal length is $|1/(4a)|$.

69. In these answers, the origin is chosen as the location of the vertex.

a. $y = ax^2$ where x and y are in feet; Since for x = æ 2.5 feet we have y = 1.25 ft., we get $1.25 = a \bullet (2.5)^2$ or a = 0.2; thus the equation is $y = 0.2x^2$

b. Focal length = $|1/(4a)| = |1/(4 \bullet 0.2)| = 1.25$ feet.

c. The horizontal line through the focal point at (0, 1.25) intersects the parabola at $x = \pm 2.5$ ft. Hence the diameter there is $2 \bullet 2.5 = 5$ ft.

Exercises for Section 8.5

71. Below is the completed table. As you can see, the average rate of change has a constant slope of -2 and thus is linear. The function to be graphed is the function y = -2x. Its graph is given to the right of the table

x	y	avg. rate of change of y wrt x	avg rate of change of avg. rate of change
- 3	- 3	n.a.	n.a.
- 2	1	4	n.a.
- 1	3	2	- 2
0	3	0	- 2
1	1	- 2	- 2
2	- 3	- 4	- 2
3	- 9	- 6	- 2

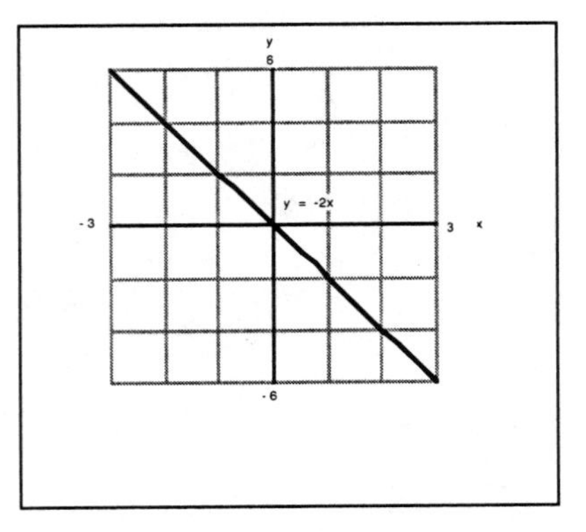

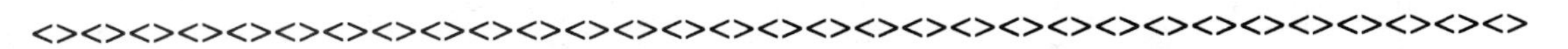

1.

Time (sec)	Distance (cm)	Avg. Vel. over previous 1/30th sec. (cm/sec)
0.0000	0.00	n.a.
0.0333	3.75	113
0.0667	8.67	147
0.1000	14.71	181
0.1333	21.77	212
0.1667	29.90	243

The average velocity over each 1/60th of a second increases rapidly as time progresses.

3. For $d = 490t^2 + 50t$:

a. 50 is measured in cm/sec; it is the initial velocity of the object falling; 490 is measured in (cm/sec)/sec and is half the acceleration due to gravity when measured in these units.

b. To the right is a small table of values.

c. Below on the left is the graph of the equation with the table points marked on it.

t	d
0.0	0.0
0.1	9.9
0.2	29.6
0.3	59.1

Graph for #3

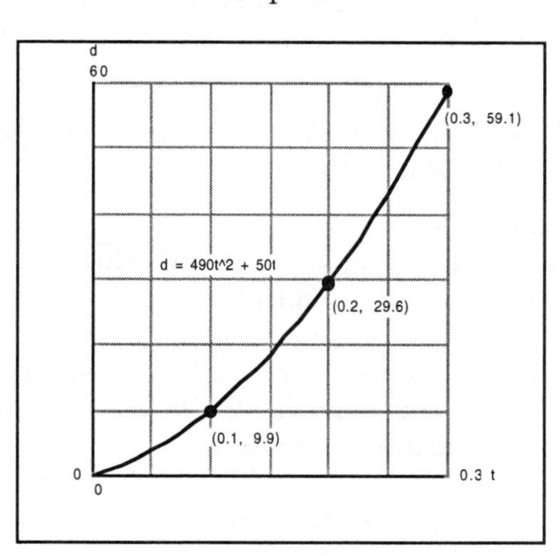

Graph for #5

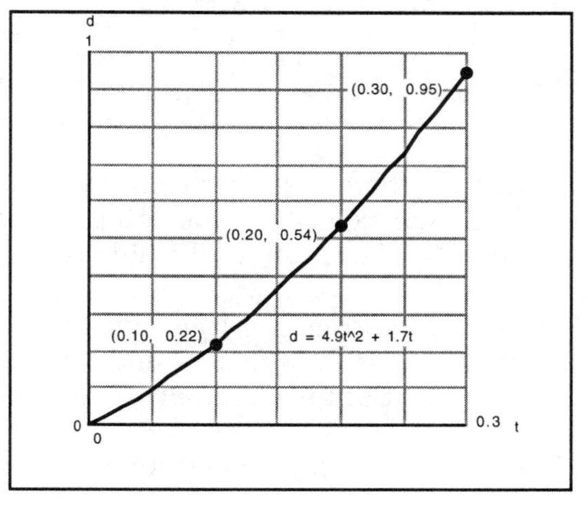

5. For $d = 4.9t^2 + 1.7t$

a. 1.7 is the initial velocity of the object falling; it is measured in meters per sec.; 4.9 is half the gravitational constant when it is measured in (meters/sec)/sec.

b. To the right is a small table of values:

c. The graph with the table points marked on it is at the bottom of the previous page on the right.

t	d
0.0	0.000
0.1	0.219
0.2	0.536
0.3	0.951

d. The results in this question are very similar to those in earlier parts of this chapter. The shape of the graph is that of a quadratic; the coefficients have the same kinds of interpretations as in earlier examples.

7. $\frac{m}{\sec^2} \bullet \sec^2 + \frac{m}{\sec} \bullet \sec = m$

9. a. $d = 490t^2 + 50t$ $\quad v(t) = 980t + 50$

b. At t = 1, d = 540 cm. and v = 1030 cm/sec. At t = 2.5, d = 3187.5 cm. and v = 2500 cm/sec.

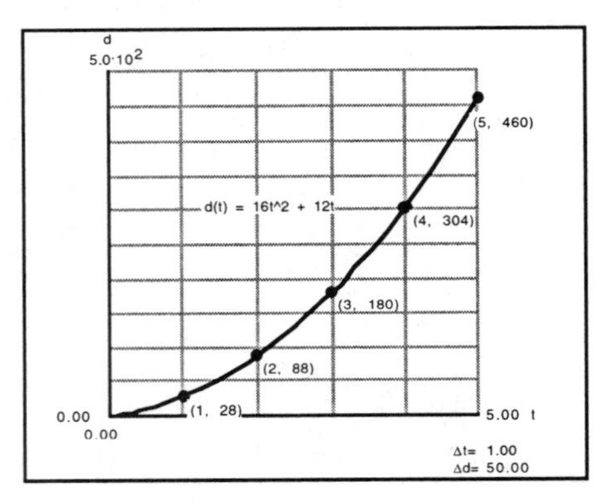

11. a. $d(t) = 16t^2 + 12t$

b.

t	d(t)
0	0
1	28
2	88
3	180
4	304
5	460

c. The graph with the plots of the points in the table of the function are given on the left.

13. The distance is measured in meters if the time is measured in seconds. The use of 4.9 for 1/2 of the gravity constant is the indicator of these units.

15. a. Answers here are expected to vary considerably.

b. Since the velocity is changing at a constant rate, a straight line should be a good fit. The graph of this line is a representation of velocity.

17. a. The coefficient of t^2 is one half the gravity constant. When distance is measured in centimeters and time in seconds, it is measured in cm/sec^2. The coefficient of t is an initial velocity and is measured in cm/sec.

b. d(0.05) = 1.59 cm. $\quad$ d(0.10) = 5.617 cm $\quad$ d(0.30) = 45.993 cm.

19. It represents an initial velocity of the object measured in meters per second.

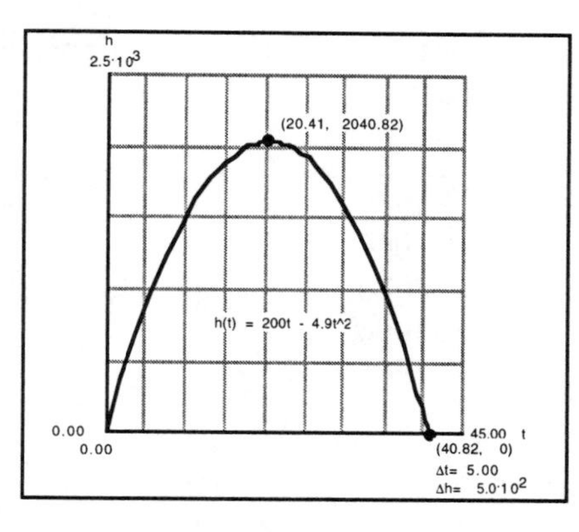

21. a. At t =0.1, h ≈ 19.951 m. At t = 2, h ≈ 380.4 m, At t = 10, h ≈ 1510 m.

b. The graph of h over t is given in the diagram to the left.

c. The object reaches a maximum height of approximately 2040.82 meters after 20.41 seconds. It reaches the ground after approximately 40.82 seconds of flight.

23. For $h = 85 - 490t^2$:

a. 85 is the height in centimeters of the falling object at the start; -490 is half the gravitational constant when measured in (cm/sec)/sec.; it is negative in value since h measures height above the ground and the gravitational constant is connected with pulling objects down. This will mean subtraction from that height.

b. The initial velocity is 0 cm/sec.
c. To the right is a table of values for this function
d. Below on the left is the graph of the function with the table entries marked on it.

t	h
0.0	85.0
0.1	80.1
0.2	65.4

Graph for **#23**

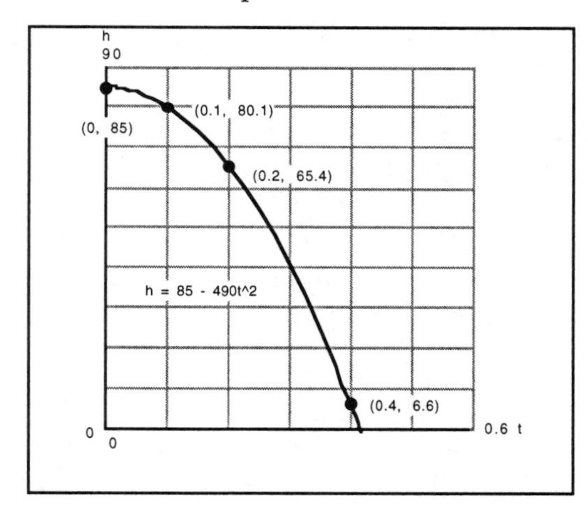

25. a. The initial velocity is positive since we are measuring height above ground and the object is going up at the start.

b. The equation of motion is $h = 50 + 10t - 16t^2$, where height is measured in feet and t in seconds.

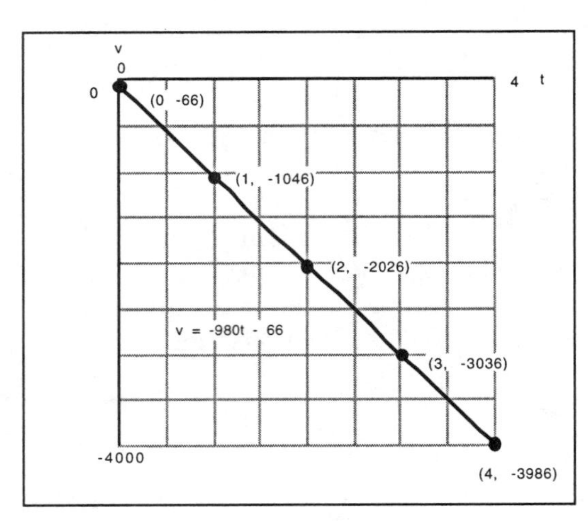

27. a. 980 cm/sec^2 since we are measuring in cm and in sec.

b.

t	v
0	-66
1	-1046
2	-2026
3	-3006
4	-3986

c. The graph is given above on the left. The object is traveling faster and faster towards the ground. The increase in downward velocity is at a constant rate, as we can see from the constant slope of the graph. This constant acceleration, of course, is due to gravity.

d. Ordinarily, if t = 0 corresponds to the actual start of the flight, then the initial condition given would indicate that the object was thrown downwards at a speed of 66 meters per second. This interpretation comes from the negative sign given to the initial velocity. But this would contradict the statement in the problem to the effect that it is a "freely falling body". In this context, another interpretation is suggested by the laboratory experiment, namely that the object started being timed at a point along its downward flight.

29. a. Its velocity starts out negative and continues to be so since the object is falling; h is measured in cm above the ground; t is measured in seconds.

b. $h = 150 - 25t - 490t^2$; for $0 \le t \le 0.528$ (the second value being the approximate time in seconds it takes for the object to hit the ground).

c. The average velocity is the slope, *i.e.*, (15-150)/0.5 = -270 cm/sec; the initial velocity is -25 cm/sec. The average velocity is 10.8 times as great in magnitude as the initial velocity.

31. Forming $\frac{d}{t} = \frac{v_0 + (v_0 + at)}{2}$ and solving for d, we get d = $\frac{2v_0 t + at^2}{2} = v_0 t + \frac{1}{2}at^2$
This is very similar in form to the falling body formula. The acceleration factor increases the velocity in a manner proportional to the square of the time traveled and the initial velocity increases the distance in a manner proportional to the time.

33. a. After 5 seconds its velocity is 110 cm/sec; after 1 minute (or 60 seconds) its velocity is 660 cm/sec; after t seconds, its velocity is: $v(t) = 60+10t$ cm/sec.

b. After 5 seconds its average velocity is $(60+110)/2 = 85$ cm/sec.

35. a. $v(t) = 200 + 60t$ meters/sec. **b.** $d(t) = 200t + 30t^2$ meters

37. a. The units used are feet and seconds and thus the equation governing the waterspout is $d = -16t^2 + v_0 t$, where d is measured in feet and t in seconds and where v_0 is the sought after initial velocity. We are given that the maximum height reached is 120 ft. The maximum height is achieved at the vertex, *i.e.*, when $t = -v_0/(-32)$. Solving the equation:

$$120 = -16\left(\frac{v_o}{32}\right)^2 + v_0\left(\frac{v_o}{32}\right) = -\frac{v_o^{\,2}}{64} + 2\frac{v_o^{\,2}}{64} = \frac{v_o^{\,2}}{64}$$ for v_0 one obtains that $v_0^{\,2} = 7680$ or $v_0 = 87.64$ ft. per second.

b. $t = v_0/32 = 87.64/32 \approx 2.75$ sec.

39. a. $d_c = v_c t + a_c t^2/2$; $d_p = a_p t^2/2$. One wants to solve for the t at which $d_c = d_p$, i.e., when

$$0 = v_c t + a_c t^2/2 - a_p t^2/2 = t(v_c + a_c t/2 - a_p t/2) = t\,(v_c + [a_c/2 - a_p/2]t).$$

This occurs at $t = 0$ (when the criminal passes by the police car) and again when $t = 2v_c/(a_p - a_c)$.

b. Now $v_c = a_c t + v_c$ and $v_p = a_p t$.. One wants to solve for the t at which $v_c = v_p$, i.e., when $a_c t + v_c = a_p t$ or for $t = (a_p - a_c)/v_c$. This does not mean that the police have caught up to the criminal but rather that the police are at that moment going as fast as the criminal is and that they are starting to go faster.